Mathematics via Problems

PART 1: Algebra

MSRI Mathematical Circles Library

Mathematics via Problems

PART 1: Algebra

Arkadiy Skopenkov

Translated from Russian by Paul Zeitz and Sergei G. Shubin

MSRI
Mathematical Sciences Research Institute
Berkeley, California

AMS
AMERICAN
MATHEMATICAL
SOCIETY
Providence, Rhode Island

This work was originally published in Russian by "МЦНМО" under the title Элементы математики в задачах, © 2018. The present translation was created under license for the American Mathematical Society and is published by permission.

This volume is published with the generous support of the Simons Foundation and Tom Leighton and Bonnie Berger Leighton.

2020 *Mathematics Subject Classification*. Primary 00-01, 00A07, 11-01, 12-01, 20-01, 26-01, 40-01, 97H20, 97H30, 97H40.

For additional information and updates on this book, visit
www.ams.org/bookpages/mcl-25

Library of Congress Cataloging-in-Publication Data

Names: Skopenkov, Arkadiy, 1972– author.
Title: Mathematics via problems: Part 1: Algebra / Arkadiy Skopenkov; translated by Sergei Shubin and Paul Zeitz.
Other titles: Matematika cherez problemy. English.
Description: Providence, Rhode Island: American Mathematical Society, 2020. | Series: MSRI mathematical circles library, 1944-8074; 25 | "MSRI, Mathematical Sciences Research Institute, Berkeley, California." | Includes bibliographical references and index.
Identifiers: LCCN 2020030058 | ISBN 9781470448783 (paperback) | 9781470462888 (ebook)
Subjects: LCSH: Algebra–Problems, exercises, etc. | Mathematics–Problems, exercises, etc. | AMS: General – Instructional exposition. | General – General and miscellaneous specific topics – Problem books. | Number theory – Instructional exposition. | Field theory and polynomials – Instructional exposition. | Group theory and generalizations – Instructional exposition. | Real functions – Instructional exposition. | Sequences, series, summability – Instructional exposition. | Mathematics education – Algebra – Elementary algebra. | Mathematics education – Algebra – Equations and inequalities. | Mathematics education – Algebra – Groups, rings, fields.
Classification: LCC QA157 .S5913 2020 | DDC 512–dc23
LC record available at https://lccn.loc.gov/2020030058

Contents

Foreword

Problems, exercises, circles, and olympiads

This is a translation of Part 1 of the book *Mathematics Through Problems: From Mathematical Circles and Olympiads to the Profession*, and is part of the MSRI Mathematical Circles Library series. The other two parts, *Geometry* and *Combinatorics*, will be published in the same series soon.

The goal of the MSRI Mathematical Circles Library series is to build a body of works in English that help to spread the "math circle" culture. A *mathematical circle* is an eastern-European notion. Math circles are similar to what most Americans would call a math club for kids, but with several important distinguishing features.

First, they are *vertically integrated*: young students may interact with older students, college students, graduate students, industrial mathematicians, professors, and even world-class researchers, all in the same room. The circle is not so much a classroom as a gathering of young initiates with elder tribespeople, who pass down *folklore*.

Second, the "curriculum," such as it is, is dominated by *problems* rather than specific mathematical topics. A problem, in contrast to an exercise, is a mathematical question that one doesn't know how, at least initially, to approach. For example, "What is 3 times 5?" is an exercise for most people but a problem for a very young child. Computing 5^{34} is also an exercise, conceptually very much like the first example, certainly harder, but only in a "technical" sense. And a question like "Evaluate $\int_2^7 e^{5x} \sin 3x \, dx$" is also an exercise for calculus students—a matter of "merely" knowing the right algorithm and how to apply it.

Problems, by contrast, do not come with algorithms attached. By their very nature, they require *investigation*, which is both an art and a science, demanding technical skill along with focus, tenacity, and inventiveness. Math circles teach students these skills, not with formal instruction, but by having them *do math* and observe others doing math. Students learn that a problem worth solving may require not minutes but possibly hours, days, or even years of effort. They work on some of the classic folklore problems and discover how these problems can help them investigate other problems. They learn how not to give up and how to turn errors or failures into opportunities for more investigation. A child in a math circle learns to do exactly what a research mathematician does; indeed, he or she does independent research, albeit on a lower level, and often—although not always—on problems that others have already solved.

Finally, many math circles have a culture similar to a sports team, with intense camaraderie, respect for the "coach," and healthy competitiveness (managed wisely, ideally, by the leader/facilitator). The math circle culture is often complemented by a variety of problem solving contests, often called *olympiads*. A mathematical olympiad problem is, first of all, a genuine problem (at least for the contestant), and usually requires an answer which is, ideally, a well-written argument (a "proof").

Why this book, and how to use it

The Mathematical Circles Library editorial board chose to translate this book because it has an audacious goal—promised by its title—to develop mathematics through problems. This is not an original idea, nor just a Russian one. American universities have experimented for years with IBL (inquiry-based learning) and Moore-method courses, structured methods for teaching advanced mathematics through open-ended problem solving.[1]

But this massive work is an attempt to curate sequences of problems for secondary students (the stated focus is high school students, but that can be broadly interpreted) that allow them to discover and recreate much of "elementary" mathematics (number theory, polynomials, inequalities, calculus, geometry, combinatorics, game theory, probability) and start edging into the sophisticated world of group theory, Galois theory, etc.

The book is not possible to read from cover to cover—nor should it be. Instead, the reader is invited to start working on problems that he or she finds appealing and challenging. Many of the problems have hints and solution sketches, but not all. No reader will solve all the problems. That's not the point—it is not a contest. Furthermore, some of the problems are not supposed to be solved, but should rather be pondered. For example, when learning about primes, it is natural to wonder whether there is always a prime between n and $2n$. Indeed, this is problem 1.6.9 (c)—the very nontrivial result known as Bertrand's postulate—and the text provides references for learning more about it. Just because it is "too advanced" doesn't mean that it shouldn't be thought about! In fact, sometimes the reader is explicitly directed to jump ahead, with references to material that appears later in the book (the authors assure the reader that this will not lead to circular reasoning).

Indeed, this is the philosophy of the book: Mathematics is not a sequential discipline, where one is presented with a definition that leads to a lemma which leads to a theorem which leads to a proof. Instead it is an adventure

---

[1]See, for example, https://en.wikipedia.org/wiki/Moore_method and http://www.jiblm.org.

filled with exciting side trips as well as wild goose chases. The adventure is its own reward, but it also, fortuitously, leads to deep understanding and appreciation of mathematical ideas that cannot be accomplished by passive reading.

English-language references

Most of the references cited in this book are in Russian. However, there are many excellent books in English (some translated from Russian). Here is a very brief list, organized by topic and chapter.[2]

Articles from *Kvant*: This superb journal is published in Russian. However, it has been sporadically translated into English under the name *Quantum*, and there are several excellent collections in English; see [**FT07, Tab99, Tab01**].

Problem collections: *The USSR Olympiad Problem Book* [**SC93**] is a classic collection of carefully discussed problems. Additionally, [**FK91**] and [**FBKY11a, FBKY11b**] are good collections of olympiads from Leningrad and Moscow, respectively. See also the nicely curated collections of fairly elementary Hungarian contest problems [**Kur63a, Kur63b, Liu01**] and the more advanced (undergraduate-level) Putnam Exam problems [**KPV02**].

Inequalities: See [**Ste04**] for a comprehensive guide and [**AS16b**] for a more elementary text. The author also recommends two classic books, [**HLP67**] and [**BB65**], and the more specialized text [**MO09**], but cautions that these are all rather advanced.

Geometry: *Geometry Revisited* [**CG67**] is a classic, and [**Che16**] is a recent and very comprehensive guide to "olympiad geometry."

Polynomials and theory of equations: See [**Bar03**] for an elementary guide, and [**Bew06**] for a historically motivated exposition of constructability and solvability and unsolvability. In Chapter 8, see the book [**Gin07**] for English translations of the *Kvant* articles [**Gin72, Gin76**], and [**Skoa**] for an abridged English version of [**Sko10**].

Combinatorics: The best book in English, and possibly any language, is *Concrete Mathematics* [**GKP94**].

Functions, limits, complex numbers, and calculus: The classic *Problems and Theorems in Analysis* by Pólya and Szegő [**PS04**] is—like the current text—a curated selection of problems but at a much higher mathematical level.

Paul Zeitz
April 2019

---

[2]We omit any supplementary Russian-language references mentioned in the original text that were not actually cited in the text.

Introduction

What this book is about and who it is for

A deep understanding of mathematics is useful both for mathematicians and for high-tech professionals. In particular, the "profession" in the title of this book does not necessarily mean the profession of mathematics.

This book is intended for high school students and undergraduates (in particular, those interested in olympiads). For more details, see "Olympiads and mathematics" on p. xvii. The book can be used both for self-study and for teaching.

This book attempts to build a bridge (by showing that there is no gap) between ordinary high school exercises and the more sophisticated, intricate, and abstract concepts in mathematics. The focus is on engaging a wide audience of students to think creatively in applying techniques and strategies to problems motivated by "real world or real work" [**Mey**]. Students are encouraged to express their ideas, conjectures, and conclusions in writing. Our goal is to help students develop a host of new mathematical tools and strategies that will be useful beyond the classroom and in a number of disciplines (cf. [**IBL**, **Mey**, **RMP**]).

The book contains the most standard "base" material (although we expect that at least some of this material is review—that not all is being learned for the first time). But the main content of the book is more complex material. Some topics are not well known in the traditions of mathematical circles, but are useful both for mathematical education and for preparation for olympiads.

The book is based on the classes taught by the author at different times at the Independent University of Moscow, at various Moscow schools and math circles, in preparing the Russian team for the International Mathematical Olympiad, in the "Modern Mathematics" summer school, in the Kirov and Kostroma Summer Mathematical Schools, in the Moscow Olympiad School, and also in the summer Conference of the Tournament of Towns.

Much of this book is accessible to high school students with a strong interest in mathematics.[3] We provide definitions or references for material that is not standard in the school curriculum. However, many topics are difficult if you study them "from scratch." Thus, the ordering of the problems helps to provide "scaffolding." At the same time, many topics are *independent* of each other. For more details, see p. xviii, "How this book is organized".

Learning by doing problems

We subscribe to the tradition of studying mathematics by solving and discussing problems. These problems are selected so that in the process of solving them the reader (more precisely, the solver) masters the fundamentals of important ideas, both classical and modern. The main ideas are developed incrementally with olympiad-style examples—in other words, by the simplest special cases, free from technical details. In this way, we show *how you can explore and discover these ideas on your own.*

Learning by solving problems is not just a serious approach to mathematics; it also continues a venerable cultural tradition. For example, the novices in Zen monasteries study by reflecting on riddles ("koans") given to them by their mentors. (However, these riddles are rather more like paradoxes than what we consider to be problems.) See, for example, [**Suz18**]; compare with [**Pla12**, pp. 26–33]. "Math" examples include [**Arn16b**, **BSe**, **RSG+16**, **KBK08**, **Pra07b**, **PS04**, **SC93**, **Sko09**, **Vas87**, **Zvo12**], which sometimes describe not only problems but also the principles of selecting appropriate problems. For the American tradition, see [**IBL**, **Mey**, **RMP**].

Learning by solving problems is difficult, in part, because it generally does not create the *illusion* of understanding. However, one's efforts are fully rewarded by a deep understanding of the material, leading to the ability to carry out similar (and sometimes rather different) reasoning. Eventually, while working on fascinating problems, readers will be following the thought processes of the great mathematicians and may see how important concepts and theories naturally evolve. Hopefully this will help them make their own equally useful discoveries (not necessarily in math)!

Solving a problem, theoretically, requires only understanding its statement. Other facts and concepts are not needed. (Actually, useful facts and ideas will be developed while solving the problems presented in this book.) Sometimes, you may need to know things from other parts of the book as indicated in the instructions and hints. For the most important problems we provide hints, instructions, solutions, and answers, located at the end of

---

[3]Some of the material is studied in math circles and summer schools by those who are just getting acquainted with mathematics (for example, 6th graders). However, the presentation here is aimed at the reader who already has at least a minimal understanding of mathematical culture. Younger students need a different approach; see, for example, [**GIF94**].

each section. However, they should be referred to only after attempting to solve a problem.

As a rule, we present the *formulation* of a beautiful or important result (in the form of a problem) before its *proof*. In such cases, one may need to solve later problems in order to fully work out the proof. This is always explicitly mentioned in hints, and sometimes even in the text. Consequently, if you fail to solve a problem, please read on. This guideline is helpful because it simulates the typical research situation.

This book "is an attempt to demonstrate learning as *dialogue* based on solving and discussing problems" (see [**KBK15**]).

A message *By A. Ya. Kanel-Belov*

To solve difficult olympiad problems, at the very least one must have a robust knowledge of algebra (particularly algebraic transformations) and geometry. Most olympiad problems (except for the easiest ones) require "mixed" approaches; rarely is a problem resolved by applying a method or idea in its pure form. Approaching such mixed problems involves combining several "crux" problems, each of which may involve single ideas in a "pure" form. *Learning to manipulate algebraic expressions is essential. The lack of this skill among olympians often leads to ridiculous and annoying mistakes.*

Olympiads and mathematics

> To him a thinking man's job was not to deny one reality
> at the expense of the other, but to include and to connect.
>
> U. K. Le Guin. *The Dispossessed.*

Here are three common misconceptions about very worthwhile goals: the best way to prepare for a math olympiad is by solving last year's problems; the best way to learn "serious" mathematics is by reading university text-books; the best way to master any other skill is with no math at all. There is a further misconception that one cannot achieve these apparently divergent goals simultaneously. The authors share the widespread belief that these three approaches miss the point and lead to harmful side effects: students become too keen on emulation, or they study the *language* of mathematics rather than its *substance*, or they underestimate the value of robust math knowledge in other disciplines.

We believe that these three goals are not as divergent as they might seem. The foundation of mathematical education should be the *solution and discussion of problems interesting to the student, during which a student learns important mathematical facts and concepts.* This simultaneously prepares the student for math olympiads and the "serious" study of mathematics, and is good for his or her general development. Moreover, it is more effective for achieving success in any one of the three goals above.

Research problems for high school students

Many talented high school or university students are interested in solving research problems. Such problems are usually formulated as complex questions broken into incremental steps; see, e.g., [**LKT**]. The final result may even be unknown initially, appearing naturally only in the course of thinking about the problem. Working on such questions is useful in itself and is a good approximation to scientific research. Therefore it is useful if a teacher or a book can support and develop this interest.

For a description of successful examples of this activity, see, for example, projects in the Moscow Mathematical Conference of High School Students [**M**]. While most of these projects are not completely original, sometimes they can lead to new results.

How this book is organized

You should not read each page in this book, one after the other. You can choose a sequence of study that is convenient for you (or omit some topics altogether). Any section (or subsection) of the book can be used for a math circle session.

The book is divided into chapters and sections (some sections are divided into subsections), with a plan of organization outlined at the start of each section. If the material of another section is needed in a problem, you can either ignore it or look up the reference. This allows greater freedom when studying the book, but at the same time it may require careful attention.

Topics of each section are arranged approximately in order of increasing complexity. The numbers in parentheses after a topic name indicate its "relative level": 1 is the simplest, and 4 is the most difficult. The first topics (not marked with an asterisk) are basic; unless indicated otherwise, you should begin your study with them. The remaining ones (marked with an asterisk) can be returned to later; unless otherwise stated, they are independent of each other. As you read, try to *return* to old material, but at a new level. Thus you should end up studying different levels of a topic *not sequentially* but as part of a mixture of topics.

The notation used throughout the book is given on p. xx. Notation and conventions particular to a specific section are introduced at the beginning of that section. The book concludes with a subject index. The numbers in bold are the pages on which *formal definitions* of concepts are given.

Sources and literature

Each chapter ends with a bibliography that relates to the entire chapter, with sources for each topic.[4] For example, we cite the book [**GKP94**],

---

[4] *Editor's note:* In the English edition all the references are combined into one list at the end of the book.

which involves both combinatorics and algebra. In addition to sources for specialized material, we also tried to include the very best popular writing on the topics studied. We hope that this bibliography, at least as a first approximation, can guide readers through the sea of popular scientific literature in mathematics. However, the great size of this genre guarantees that many remarkable works had to be omitted. Please note that items in the bibliography are not necessary for solving the problems in this book, unless explicitly stated otherwise.

Many of the problems are not original, but the source (even if it is known) is usually not specified. When a reference is provided, it comes after the statement of the problem, so that the reader can compare his or her solution with the one given there. When we know that many problems in a section come from one source, we mention this.

We do not provide links to online versions of articles in the popular magazines *Kvant* (the English magazine *Quantum* is based on *Kvant*) and *Matematicheskoe Prosveshchenie* ("Mathematical Enlightment"); they can be found at the websites `http://kvant.ras.ru`, `http://kvant.mccme.ru`, `https://en.wikipedia.org?wiki?Quantum_Magazine`, and `http://www.mccme.ru/free-books/matpros.html`.

Acknowledgments

We are grateful for the serious work of translators and editors David Scott, Sergei Shubin, and Paul Zeitz. We thank the reviewers for helpful comments, specifically, A. V. Antropov (Chapters 1 and 2), A. I. Sgibnev (Chapters 3 and 7), S. L. Tabachnikov (Chapter 8), and A. I. Khrabrov (Chapters 5 and 6), and also the anonymous reviewers of selected materials. We thank A. I. Kanel-Belov, the author of some material in this book, who also contributed a number of useful ideas and comments. We thank our students for asking challenging questions and pointing out errors. Further acknowledgments for specific sections are given directly in the text.

We apologize for any mistakes, and will be grateful to readers for pointing them out.

Grant support

A. B. Skopenkov was partially supported by grants from the Simons Foundation and the Dynasty Foundation.

Contact information

A. B. Skopenkov: Moscow Institute of Physics and Technical (State University) and Independent University of Moscow, `https://users.mccme.ru/skopenko`.

Numbering and notation

Sections in each chapter are arranged approximately in order of increasing complexity of the material. The numbers in parentheses after the section name indicate its "relative level": 1 is the simplest, and 4 is the most difficult. The first sections (not marked with an asterisk) are basic; unless indicated otherwise, you can begin to study the chapter with them. The remaining sections (marked with an asterisk) can be returned to later; unless otherwise stated, they are independent of each other.

If a mathematical fact is formulated as a problem, then the objective is to prove this fact. Open-ended questions are called *challenges*; here one must come up with clear wording and a proof; cf., for example, [**VINK10**].

The most difficult problems are marked with asterisks (*). If the statement of the problem asks you to "find" something, then you need to give a "closed form" answer (as opposed to, say, an unevaluated sum of many terms).

Once again, if you are unable to solve a problem, continue reading: later problems may turn out to be hints.

Notation

- $\lfloor x \rfloor = [x]$ — (lower) integer part of number x ("floor"); that is, the largest integer not exceeding x.
- $\lceil x \rceil$ — upper integer part of number x ("ceiling"); that is, the smallest integer not less than x.
- $\{x\}$ — fractional part of number x; equal to $x - \lfloor x \rfloor$.

- $d|n$, or $n \vdots d$ — d *divides* n; that is, there exists an integer k such that $n = kd$ (the number d is called a *divisor* of the number n; we assume that $d \neq 0$).
- $\mathbb{R}$, $\mathbb{Q}$, and $\mathbb{Z}$ — the sets of all real numbers, rational numbers, and integers, respectively.
- $\mathbb{Z}_2$ — the set $\{0, 1\}$ of remainders upon division by 2 with the operations of addition and multiplication modulo 2.
- $\mathbb{Z}_m$ — the set $\{0, 1, \ldots, m - 1\}$ of remainders upon division by m with the operations of addition and multiplication modulo m. (Specialists in algebra often write this set as $\mathbb{Z}/m\mathbb{Z}$ and use $\mathbb{Z}_m$ for the set of *m-adic integers* for the prime m.)
- $\binom{n}{k}$ — the number of k-element subsets of an n-element set (also denoted by C_n^k).
- $|X|$ — the number of elements in set X.
- $A - B = \{x \mid x \in A \text{ and } x \notin B\}$ — the difference of the sets A and B.
- $A \sqcup B$ — the disjoint union of the sets A and B; that is, the union of A and B viewed as the union of disjoint sets.

- $A \subset B$ — means the set A is contained in the set B. In some books, this is denoted by $A \subseteq B$, and $A \subset B$ means "the set A is in the set B and is not equal to B."
- We abbreviate the phrase "Define x to be a" to $x := a$.

Chapter 1

Divisibility

The parts of this chapter used in the rest of the book are: the Euclidean algorithm and its applications (problems 1.5.7 and 1.5.9), the language of congruences (section 4, "Division with a remainder and congruences"), and some simple facts (e.g., problem 1.1.3 and 1.3.2).

In this chapter all variables are integers. Many solutions are based on M. A. Prasolov's texts.

1. Divisibility (1)

1.1.1. (a) State and prove the rules of divisibility by 2, 4, 5, 10, 3, 9, 11.

(b) Is the number $11\ldots1$ consisting of 1993 ones divisible by 111111?

(c) Prove that the number $1\ldots1$ consisting of 2001 ones is divisible by 37.

1.1.2. If a is divisible by 2 and not divisible by 4, then the number of even divisors of a is equal to the number of its odd divisors.

1.1.3. Which of the following statements are correct for any a and b? (Recall the notation $a|b$ defined on p. xx.)

(a) $2|(a^2 - a)$.

(b) $4|(a^4 - a)$.

(c) $6|(a^3 - a)$.

(d) $30|(a^5 - a)$.

(e) If $c|a$ and $c|b$, then $c|(a + b)$.

(f) If $b|a$, then $bc|ac$.

(g) If $bc|ac$ for some $c \neq 0$, then $b|a$.

To solve problem 1.1.3 (c), we used 1.1.4 (a). Prove it using the definition of divisibility, but not using the Unique Factorization Theorem (problem 1.2.8 (d))! The use of this theorem might lead to a circular argument since a result similar to 1.1.4 (a) is usually used in a proof of uniqueness of factorization.

1.1.4. (a) If a is divisible by 2 and 3, then it is also divisible by 6;
(b) If a is divisible by 2, 3, and 5, then it is also divisible by 30;
(c) If a is divisible by 17 and 19, then it is also divisible by 323.

1.1.5. (a) If k is not divisible by 2, 3, or 5, then $k^4 - 1$ is divisible by 240.
(b) If $a + b + c$ is divisible by 6, then $a^3 + b^3 + c^3$ is also divisible by 6.
(c) If $a + b + c$ is divisible by 30, then $a^5 + b^5 + c^5$ is also divisible by 30.
(d) If $n \geq 0$ then $20^{2n} + 16^{2n} - 3^{2n} - 1$ is divisible by 323.

Suggestions, solutions, and answers

1.1.1. In the proofs of divisibility rules below, we denote the number in the statements by $n = \pm(10^m a_m + 10^{m-1} a_{m-1} + \ldots + 10 a_1 + a_0)$ for some $0 \leq a_i \leq 9$.

Rule of divisibility by 2: An integer is divisible by 2 if and only if the last digit of the integer is divisible by 2.

Proof. Clearly, the number $n - a_0$ is even. Suppose a_0 is also even. If a number divides each term of the sum, it divides the sum. Therefore n is even. Conversely, if a number n is even, then a_0 is even. ☐

Rule of divisibility by 4: An integer n is divisible by 4 if and only if the number formed by its last two digits is divisible by 4.

Proof. Clearly, the number $(n - 10 a_1 - a_0)$ is divisible by 4. Suppose that the number $a_0 + 10 a_1$ formed by the last two digits of n is divisible by 4. Then n is divisible by 4. Conversely, if $4 | n$ then $4 | (a_0 + 10 a_1)$. ☐

Rule of divisibility by 5: An integer is divisible by 5 if and only if its last digit is 5 or 0.
Prove this similarly to proving the rule of divisibility by 2.

Rule of divisibility by 10: An integer is divisible by 10 if and only if its last digit is 0.
Prove this similarly to proving the rule of divisibility by 2.

Rule of divisibility by 3: An integer n is divisible by 3 if and only if the sum of its digits is divisible by 3.

Proof. Subtract the sum of digits from the number and group the summands as follows:

$$n - a_m - a_{m-1} - \ldots - a_1 - a_0$$
$$= (10^m - 1)a_m + (10^{m-1} - 1)a_{m-1} + \cdots + (10 - 1)a_1 + (1 - 1)a_0.$$

The number $10^k - 1 = (10 - 1)(10^{k-1} + 10^{k-2} + \ldots + 10 + 1)$ is divisible by 3. The rule of divisibility by 3 follows from this observation. $\square$

Rule of divisibility by 9: An integer n is divisible by 9 if and only if the sum of its digits is divisible by 9.

Prove this similarly to proving of the rule of divisibility by 3.

Rule of divisibility by 11: Subtract the sum of all digits of n at odd positions from the sum of all digits at even positions. The number n is divisible by 11 if and only if the resulting number $f(n)$ is divisible by 11.

Proof. First, we will prove that for any $m \geq 0$ the number $10^m - (-1)^m$ is divisible by 11. For odd m, the number $10^m + 1 = (10+1)(10^{m-1} - 10^{m-2} + 10^{m-3} - \ldots - 10 + 1)$ is divisible by 11. For even m, the number $10^m - 1$ is divisible by $10^2 - 1$ and hence divisible by 11. Now we have

$$n - f(n) = (10^m - (-1)^m)a_m + (10^{m-1} - (-1)^{m-1})a_{m-1}$$
$$+ \ldots + (10 + 1)a_1 + (1 - 1)a_0.$$

Since every term of the sum on the right-hand side of the equation is divisible by 11, n is divisible by 11 if and only if $f(n)$ is divisible by 11. $\square$

1.1.3. *Answers*: (a, c, d, e, f) true; (b) false.

(a) We have $a^2 - a = a(a - 1)$. Taken in the natural order, every other integer is even; thus one of the numbers a or $a - 1$ is even, so their product $a^2 - a$ is also even.

(b) 4 does not divide $(2^4 - 2) = 14$.

(c) We have $a^3 - a = a(a - 1)(a + 1)$. The number $a(a - 1)$ is divisible by 2 while $(a - 1)a(a + 1)$ is divisible by 3. Thus $a^3 - a$ is divisible by 2 and 3, and, as follows from 1.1.4 (a), it is divisible by 6.

(d) We have $a^5 - a = a(a - 1)(a + 1)(a^2 + 1)$. Now, $a(a - 1)$ is divisible by 2 while $(a - 1)a(a + 1)$ is divisible by 3. If none of the numbers $a - 1$, a, and $a + 1$ is divisible by 5, then the remainder from dividing a by 5 is equal to 2 or 3. Thus $a^2 + 1$ is divisible by 5. Then, as follows from 1.1.4 (b), $a^5 - a$ is divisible by 30.

(e) If $a = kc$ and $b = mc$, then $a + b = (k + m)c$.

(f) If $a = kb$ then $ac = k(bc)$.

(g) If $ac = kbc$ then $c(a - kb) = 0$. Since $bc \neq 0$ we have $c \neq 0$; therefore $a = kb$.

1.1.4. (a) *Hint.* We have $3a - 2a = a$.

Solution. Since $2|a$ we have $6|3a$, and since $3|a$ we have $6|2a$; therefore $6|(3a - 2a) = a$.

(b) *Hint.* $6a - 5a = a$.

Solution. From the given conditions and part (a) above we have $6|a$ and $5|a$. Therefore $30|6a$ and $30|5a$, so $30|(6a - 5a) = a$.

(c) *Hint.* $19a - 17a = 2a$, $17a - 8 \cdot 2a = a$.

Solution. From the given conditions we have $17|a$ and $19|a$. Therefore $17 \cdot 19|17a$ and $19 \cdot 17|19a$. So $17 \cdot 19|(19a - 17a) = 2a$. Then $17 \cdot 19|(17a - 8 \cdot 2a) = a$.

1.1.5. (d) The number $(a^n - b^n) = (a - b)(a^{n-1} + a^{n-2}b + \ldots + b^{n-1})$ is divisible by $(a - b)$. Therefore $20^{2n} + 16^{2n} - 3^{2n} - 1 = (20^{2n} - 3^{2n}) + ((16^2)^n - (1^2)^n)$ is divisible by 17. Similarly, $20^{2n} + 16^{2n} - 3^{2n} - 1 = (20^{2n} - 1) + ((16^2)^n - (3^2)^n)$ is divisible by 19. Then, according to 1.1.4 (c), $20^{2n} + 16^{2n} - 3^{2n} - 1$ is divisible by 323.

2. Prime numbers (1)

An integer $p > 1$ is said to be a *prime* if it does not have positive divisors other than p and 1. An integer q is a *composite* if it has at least one positive divisor different from 1 and $|q|$. (Thus 1 is neither a prime nor a composite number.)

1.2.1. (a) **Lemma.** If $a¿1$ is not divisible by any prime $p \leq \sqrt{a}$, then a is a prime.

(b) **Sieve of Eratosthenes.** Let $p_1, \ldots, p_k$ all be primes between 1 and n. For each $i = 1, \ldots, k$ we will cross out all numbers between 1 and n^2 which are divisible by p_i. Numbers which are left are all primes between n and n^2.

(c) Write down all primes between 1 and 200.

1.2.2. (a) Find all p such that p, $p + 2$, and $p + 4$ are primes.

(b) Prove that if the number $11\ldots1$ consisting of n ones is a prime, then n is a prime.

(c) Prove that the converse of (b) is not true.

Theorem 1.2.3 (Euclid). (a) There are infinitely many primes.

(b) There are infinitely many primes of the form $4k + 3$.

Compare to problem 2.3.3 (f). Using advanced techniques it's possible to prove the following statement.

Theorem 1.2.4 (Dirichlet). If the integers $a, b > 0$ have no common divisors other than ± 1, then there are infinitely many primes of the form $ak + b$.

1.2.5. Let p_n denote the nth prime number (in ascending order).
(a) Prove that $p_{n+1} \leq p_1 \cdot \ldots \cdot p_n + 1$
(b) Prove that $p_{n+1} \leq p_1 \cdot \ldots \cdot p_n - 1$ for $n \geq 2$.
(c)* Prove that there is a perfect square between $p_1 + \ldots + p_n$ and $p_1 + \ldots + p_{n+1}$.

1.2.6. (a) Is it true that for any n, the number $n^2 + n + 41$ is a prime?
(b) Prove that for any non-constant quadratic function f with integer coefficients, there exists an integer n such that the number $|f(n)|$ is composite.
(c) Prove that for any non-constant polynomial f with integer coefficients, there exists an integer n such that the number $|f(n)|$ is composite.

1.2.7. There exist 1000 consecutive numbers, none of which is
(a) a prime;
(b) a power of a prime.

1.2.8. (a) Any positive integer may be decomposed into a product of prime numbers.
(b) An even number is called *primish* if it is not a product of two smaller positive even numbers. Is the decomposition of an even number into a product of primish numbers necessarily unique? (See a more meaningful example in problem 3.7.3 (b).)
(c)* If a number is equal to the product of two primes, this decomposition is unique up to the order of the factors.
(d) **Fundamental Theorem of Arithmetic.** The decomposition of any positive integer into a product of primes is unique up to the order of the factors. (This theorem is often referred to as the Unique Factorization Theorem or the Canonical Decomposition Theorem.)

For the (usual) solution of (b) and (c) you will need the lemmas in problem 1.5.7. See also problem 3.4.5.

Suggestions, solutions, and answers

1.2.2. (a)*Answer*: $p = 3$.
Solution. The numbers p, $p + 2$, and $p + 4$ have different remainders upon division by 3. Therefore one of them is divisible by 3. This number is a prime, so it is equal to 3. Since all primes by definition are positive

integers, then $p+4 \neq 3$. Since 1 is not a prime, $p+2 \neq 3$. Thus $p = 3$. This is indeed our solution, because 3, 5, and 7 are primes.

(b) Assume to the contrary that n is composite, i.e., $n = ab$, where $a, b > 1$. We have $x^b - 1 = (x - 1)(x^{b-1} + x^{b-2} + \ldots + x + 1)$. Substituting $x = 10^a$ we see that $11\ldots1 = \frac{10^n - 1}{9}$ is divisible by $\frac{10^a - 1}{9}$.

(c) The converse statement is false: $111 = 37 \cdot 3$.

1.2.7. (a) For example, $1000!+2$, $1000!+3, \ldots, 1000!+1001$. The problem can also be solved similarly to part (b).

(b) Take different primes $p_1, p_2, \ldots, p_{2000}$. The *Chinese Remainder Theorem* 1.5.10 (d) implies that there exists n such that $n + i$ is divisible by $p_{2i-1}p_{2i}$ for any $i = 1, 2, \ldots, 1000$.

1.2.8. (a) Suppose that not every integer is a product of primes. Consider the smallest positive integer n which is not a product of primes. If it is not a prime, then it is a composite number, so $n = ab$ for some $a, b > 1$. Therefore $n > a$ and $n > b$. But n is the smallest integer not equal to a product of primes, so a and b are both products of primes. Hence n is also a product of primes. This contradicts our assumption.

(d) Suppose the assertion is false. Consider the smallest number n having two different canonical decompositions: $n = p_1^{a_1} \cdot p_2^{a_2} \cdot \ldots \cdot p_m^{a_m} = q_1^{b_1} \cdot q_2^{b_2} \cdot \ldots \cdot q_k^{b_k}$. Since n is minimal, none of the numbers p_i is equal to any q_j, for otherwise we could divide both sides of the equality by this number and get a smaller number with two different canonical decompositions. On the other hand, q_1 divides $p_1^{a_1} \cdot p_2^{a_2} \cdot \ldots \cdot p_m^{a_m}$ and therefore, as follows from 1.5.7 (c), q_1 divides one of numbers p_i. Since p_i is a prime, we have $q_1 = p_i$. This contradicts our assumption.

3. Greatest common divisor (GCD) and least common multiple (LCM) (1)

The integers a and b are said to be *relatively prime* if they don't have common divisors other than ±1.

An integer is said to be the *greatest common divisor* (GCD) of two positive integers a and b if it is the greatest number that divides both a and b. We denote the GCD of a and b by (a, b) or $\text{GCD}(a, b)$ or $\gcd(a, b)$.

1.3.1. Find all possible values:
(a) $(n, 12)$; (b) $(n, n+1)$; (c) $(n, n+6)$; (d) $(2n+3, 7n+6)$; (e) $(n^2, n+1)$.

Lemma 1.3.2. For $a \neq b$ the following equality is valid: $(a, b) = (|a - b|, b)$.

1.3.3. (a) $(a, b) = b$ if and only if a is divisible by b.

(b) The numbers $\frac{a}{(a,b)}$ and $\frac{b}{(a,b)}$ are relatively prime.

(c)* The number (a, b) is divisible by any common divisor of a and b.

(d)* We have $(ca, cb) = c(a, b)$ for any $c > 0$.

To solve problems marked with an asterisk, you will need the lemmas in 1.5.7.

1.3.4. (a) For all positive m and n we have

$$(2m, 2n) = 2(m, n), \quad (2m + 1, 2n) = (2m + 1, n),$$

$$(2m + 1, 2n + 1) = (2m + 1, m - n) \quad \text{for } m > n.$$

(b) *Binary algorithm.* Using the equalities from (a) construct an algorithm for finding the GCD.

1.3.5.* If a fraction $\frac{a}{b}$ is irreducible, then the fraction $\frac{a+b}{ab}$ is also irreducible.

An integer is said to be the *least common multiple* (LCM) of two positive integers a and b if it is the smallest number that is divisible by a and b. We denote the LCM of a and b by $[a, b]$ or LCM(a, b) or lcm(a, b).

1.3.6. Find $[192, 270]$.

1.3.7. (a) $[a, b] = a$ if and only if a is divisible by b.

(b) The numbers $\frac{[a,b]}{a}$ and $\frac{[a,b]}{b}$ are relatively prime.

(c)* Any common multiple of a and b is divisible by $[a, b]$.

(d)* $[ca, cb] = c[a, b]$ for any $c > 0$.

Suggestions, solutions, and answers

1.3.1. *Answers*: (a) 1,2,3,4,6,12. (b) 1. (c) 1,2,3,6. (d) 1,3,9. (e) 1.

Solutions.

(a) The number $(12, n)$ is a positive divisor of 12. Let $d|12$. The number d does not have divisors greater than itself, so $(12, d) = d$. Thus, all positive divisors of 12 satisfy the condition of the problem.

(b) Let $d|n, d|(n + 1)$, and $d > 0$. Then $d|(n + 1 - n) = 1$, so $d = 1$.

(c) By Lemma 1.3.2 above, $(n, n + 6) = (6, n)$. Similarly to (a), all positive divisors of 6 satisfy the condition of the problem.

(d) By Lemma 1.3.2, $(2n + 3, 7n + 6) = (2n + 3, 5n + 3) = (2n + 3, 3n) = (2n + 3, n - 3) = (n + 6, n - 3) = (n + 6, 9)$.

Thus, all positive divisors of 9 satisfy the condition of the problem.

(e) Let $d > 0$ be a common divisor of the numbers $n + 1$ and n^2. Thus $d|(n + 1)(n - 1) = n^2 - 1$ by Lemma 1.3.2. So $d|(n^2 - (n^2 - 1)) = 1$, and hence $d = \pm 1$.

1.3.2. The statement follows from the fact that the set of common divisors of a and b coincides with the set of common divisors of a and $a \pm b$. Indeed, if $d|a$ and $d|b$ then $d|(a \pm b)$. Conversely, if $d|(a \pm b)$ and $d|a$ then $d|(a \pm b - a) = \pm b$.

1.3.3. (a) Let $b|a$. Since any positive divisor of a nonzero integer n does not exceed $|n|$, we have $(a, b) = |b|$. Conversely, let $(a, b) = |b|$. Then $b|a$ by definition.

(b) If $d > 0$ is a common divisor of $\frac{a}{(a,b)}$ and $\frac{b}{(a,b)}$, then $d \cdot (a, b)$ is a common divisor of a and b. If $d > 1$ this is a contradiction.

(c) Let $a > b \geq 0$. In the proof of Lemma 1.3.2 we showed that the set of common divisors of a and b coincides with the set of common divisors of a and $a \pm b$. Apply the Euclidean algorithm to the pair of numbers $a_0 = a$ and $b_0 = b$ (see problem 1.5.9 (b)). The numbers a_k and b_k obtained in the kth step are positive. The common divisors of a_k and b_k coincide with common divisors of $a_k - b_k$ and b_k. Therefore all common divisors (and, in particular, the GCD) of all intermediate pairs are the same. At the final step of the Euclidean algorithm, we see that divisors of the number $d = \gcd(a, b)$ coincide with common divisors of the numbers a and b.

(d) The number $c(a, b)$ is a common divisor of the numbers ca and cb.

To prove this we show that $(ca, cb)|c(a, b)$. Obviously $c|ca$ and $c|cb$. From (c) above we conclude that $c|(ca, cb)$, so $(ca, cb) = ck$ for some integer k. The GCD of two numbers divides each of them, so $(ck)|(ca)$ and $(ck)|(cb)$. Thus $k|a$ and $k|b$. From (c) it follows that $k|(a, b)$. Multiplying both sides by c, we see that $(ca, cb)|c(a, b)$.

4. Division with remainder and congruences (1)

Theorem 1.4.1 (Division with a remainder). (a) For any a and $b \neq 0$ there exists q such that $q|b| \leq a < (q + 1)|b|$.

(b) For any a and $b \neq 0$ there exist unique q and r such that $a = bq + r$ and $0 \leq r < |b|$. The number q is said to be the *quotient* and the number r is said to be the *remainder* of division of a by b.

1.4.2. (a, b, c) Find the quotients and remainders for

(a) 1996 divided by -17;

(b) -17 divided by 4;

(c) $n^2 + n + 1$ divided by $n + 1$, for any n.

(d) Find all possible quotients and all possible remainders when dividing 57 by some number. (More precisely, assume that $57 = bq + r$ is division with remainder. Find the list of all possible q's and the list of all possible r's.)

Hint. There is a quicker way to do this than dividing 57 by $1, 2, 3, \ldots$, listing all resulting pairs (q, r), and removing identical entries.

1.4.3. Find
  (a) the remainder upon dividing 3^{16} by 23;
  (b) the last digit of the number $1997^{1997^{1997}}$.

To solve the problem above (among others), it's useful to be familiar with the following notion: The integers a and b are said to be *congruent modulo* $m \neq 0$ if $a - b$ is divisible by m (or, equivalently, if a and b have equal remainders upon division by m). This is denoted by $a \equiv b \pmod{m}$, or $a \equiv b \bmod m$, or $a \equiv b \ (m)$, or $a \underset{m}{\equiv} b$.

1.4.4. Properties of congruences: For any $a, b, m \neq 0$ the following statements are true:
  (a) Transitivity: If $a \equiv b \ (m)$ and $b \equiv c \ (m)$, then $a \equiv c \ (m)$.
  (b) Addition: If $a \equiv b \ (m)$ and $c \equiv d \ (m)$, then $a + c \equiv b + d \ (m)$.
  (c) Multiplication: If $a \equiv b \ (m)$ and $c \equiv d \ (m)$, then $ac \equiv bd \ (m)$.
  (d) Multiplication by an integer: If $a \equiv b \ (m)$, then $ac \equiv bc \ (mc)$ for any $c \neq 0$.
  (e)* Division by an integer: If $ac \equiv bc \ (m)$ and $(m, c) = 1$, then $a \equiv b \ (m)$.

1.4.5. (a) Any number is congruent mod 9 and mod 3 to the sum of its digits.
  (b) Formulate and prove similar rules of divisibility for 2, 4, and 11.

1.4.6. The sequence of remainders of a^n $(n = 0, 1, \ldots)$ upon division by $b \neq 0$ becomes periodic starting from some n.

Hints

1.4.1. (a) Use induction on a going "up" and "down." The base case when $0 \leq a \leq |b|$ is obvious. If $a \geq |b|$, then the inductive step reduces the assertion to the statement about $a - b$. If $a < 0$, then the next step reduces the assertion to the statement for $a + |b|$.
  (b) This statement is equivalent to (a).

1.4.3. We have

$$3^{16} = (3^2)^8 = 9^8 = (9^2)^4 = 81^4 \equiv 12^4 = (12^2)^2 \equiv 6^2 \equiv 13 \bmod 23.$$

5. Linear Diophantine equations (2)

1.5.1. (a) A grasshopper moves along a line jumping 6 cm or 10 cm in either direction. What points can it get to?

(b) On the island of Utopia, each week consists of 7 days, and each month has 31 days. Sir Thomas Moore lived there for 365 days. Was one of the days necessarily Friday the 13th?

(c) Mike added together the day of his birth multiplied by 12 and the number of the month of his birth multiplied by 31 and got 670. What is his birthday? Find all possible solutions!

(d) Solve the equation $nx + (2n - 1)y = 3$, where n is a given number (from here on we mean to find a solution in integers).

1.5.2. (a) One can make change for any amount of money greater than 23 yuan using just 5- and 7-yuan coins.

(b)* Find the smallest number m such that one can make change for any amount of money greater than m yuan using 12-, 21-, and 28-yuan coins.

1.5.3. A cue ball is launched from the corner of a billiard table at angle $45°$. Will the ball hit the pin standing at the point $(2, 1)$, if the table is a rectangle with one of its vertices at the origin of the coordinate plane and another one at the point

(a) $(12, 18)$;   (b) $(13, 18)$?

1.5.4. The equation $19x + 17y = 1$ has a solution in integers.

1.5.5. Let a and b be integers that are not both equal to 0 and let $c \in \mathbb{Z}$.

(a) **Theorem.** Let both a and b be nonzero. If a pair (x_0, y_0) is a solution of $ax + by = c$, then the set of all solutions of the equation is

$$\left\{ \left(x_0 + \frac{b}{(|a|, |b|)} t, \ y_0 - \frac{a}{(|a|, |b|)} t \right) \middle| \ t \in \mathbb{Z} \right\}.$$

(b) The equation $ax + by = c$ has a solution if and only if the equation $(a - b)u + bv = c$ has a solution.

(c) **Theorem.** The equation $ax + by = c$ has a solution if and only if c is divisible by (a, b).

(d) Construct an algorithm that either finds at least one solution of the equation $ax + by = c$ or reports that there are no solutions.

1.5.6. For any a and b not equal to 0 simultaneously, let $M = \{ax + by \mid x, y \in \mathbb{Z}\}$.

(a) Any element of M is divisible by the smallest positive element of M.

(b) The smallest positive element of M is equal to (a, b).

1.5.7. Let a and b be integers that are not both equal to 0 and let $c \in \mathbb{Z}$.

(a) **GCD representation lemma.** There exist x and y such that $ax + by = (a, b)$.

(b) **Lemma.** If $(b, c) = 1$ and $c|ab$, then $c|a$.

(c) **Euclid's lemma.** If p is a prime and $p|ab$, then $p|a$ or $p|b$.

(d) **Lemma.** If $(b, c) = 1$, $b|a$, and $c|a$, then $bc|a$.

1.5.8. (a) Find $(2^{91} - 1, 2^{63} - 1)$.

(b) Find $(2^{2^k} + 1, 2^{2^l} + 1)$.

(c) For which a, b, and n is $n^a + 1$ divisible by $n^b - 1$?

1.5.9. (a) For any a and $b \neq 0$ we have the equality $\gcd(a, b) = \gcd(b, r)$, where r is the remainder on division of a by b.

(b) For a pair of numbers $(a_0, b_0) \neq (0, 0)$, the *Euclidean algorithm* constructs the sequence of pairs (a_k, b_k) by the following rules:

- If $b_k = 0$, set $d := a_k$ and halt the algorithm.
- If $b_k \neq 0$, set $a_{k+1} := b_k$ and let b_{k+1} be equal to the remainder when a_k is divided by b_k.

Prove that for any pair of numbers $(a_0, b_0) \neq (0, 0)$, the Euclidian algorithm will come to an end and return $d = \gcd(a_0, b_0)$.

1.5.10. Solve the following systems of congruences:

(a) $\begin{cases} x \equiv -1 \ (7), \\ x \equiv 15 \ (5); \end{cases}$ (b) $\begin{cases} x \equiv 6 \ (12), \\ x \equiv 8 \ (20); \end{cases}$ (c) $\begin{cases} x \equiv 7 \ (8), \\ x \equiv 18 \ (25), \\ 6x \equiv 2 \ (7). \end{cases}$

(d) **The Chinese Remainder Theorem.** If nonzero integers $m_1, \ldots,$ m_s are pairwise relatively prime, then for any integers $a_1, \ldots, a_s$, there exists x such that $x \equiv a_i \ (m_i)$ for all $i = 1, \ldots, s$.

(e) Construct an algorithm for finding x.

Suggestions, solutions, and answers

1.5.1. (a) *Answer*: The grasshopper can get to all points whose distances from the starting point are even.

Solution. The grasshopper jumps even distances, so it can move away from the starting point only by an even distance. To show that it can get to the point located at a distance $2n$ to the right of the starting point, make $2n$ jumps by 6 to the right and n jumps by 10 to the left, since $6(2n) - 10n = 2n$. An analogous argument works for points located to the left of the starting point.

(b) Consider 7 consecutive months during which Sir Thomas Moore was on the island, numbered 1 to 7 in the same way as we number days of the week. The number of days in a month has the remainder 3 upon division by 7. This means that if the 13th day of the ith month is the kth day of the week, then the 13th day of the $(i+1)$th month will be the $(k+3)$th day of the week modulo 7. Therefore, the days of the week of 13th days of the seven months are $k, k+3, k+6, k+2, k+5, k+1, k+4$ modulo 7. This contains all 7 days of the week among them. Thus, one of them will be Friday.

1.5.2. (a) If $24 \leq n < 29$, we can make change for n yuan as follows:

$$24 = 2 \cdot 5 + 2 \cdot 7, \quad 25 = 5 \cdot 5, \quad 26 = 5 + 3 \cdot 7, \quad 27 = 4 \cdot 5 + 7, \quad 28 = 4 \cdot 7.$$

We will prove the problem's assertion by induction on n. We just proved it for $24 \leq n < 29$. If $n \geq 29$, by the induction hypothesis, we can make change for $n - 5$ yuan using 5- and 7-yuan coins.

1.5.5. (c) Assume that $a \geq b > 0$ and use induction on $a + b$.

1.5.7. (a) The statement follows from 1.5.5 (c), or from 1.5.6 (b) (or can be proved similarly).
(b) Use part (a).
(c) Use part (b).
Another hint. For fixed numbers p and $a \geq 0$, find the smallest positive number b satisfying the following conditions: $p|ab$ and b is not divisible by p. It's clear that if $p|ab$, then $p|a(b-p)$. Therefore the minimality of b implies that $b \leq p$. Since $p|ab$, we have $ab \geq p$. Consider integers $b, 2b, \ldots, (a-1)b, ab$. Among them there is an integer i satisfying $(i-1)b < p \leq ib$. If $p = ib$, then $b = 1$, so $p|ab$. Now let $p \leq ib$. Note that $0 \leq ib-p \leq b$ and $p|a(ib-p)$. This contradicts the minimality of b.

1.5.8. (a) Prove that $(n^a - 1, n^b - 1) = n^{(a,b)} - 1$.

1.5.9. (b) If $b_k \neq 0$, then for any two consecutive steps, the largest numbers in a pair will decrease. So at some step the largest number in the pair will reach its minimal value and the algorithm will halt. Therefore at some step we will obtain the pair $(a_k, 0)$. Consequently, $a_k = \gcd(a_k, 0) = \gcd(a_0, b_0)$.

1.5.10. *Answers:* (a) $x \equiv 20\ (35)$;   (b) $\emptyset$ (empty set);   (c) $x \equiv 943\ (1400)$.

6. Canonical decomposition (2*)

The existence of prime factorization (problem 1.2.8 (a)) implies that for any number $n \geq 2$, there are distinct primes $p_1, \ldots, p_m$ and positive integers $\alpha_1, \ldots, \alpha_m$ such that $n = p_1^{\alpha_1} \cdot \ldots \cdot p_m^{\alpha_m}$. This representation is said to be the *canonical decomposition* of the number n. It is uniquely determined up to the order of the factors (problem 1.2.8 (d)).

1.6.1. Find the canonical decomposition of the following numbers:
(a) 1995;   (b) 17!;   (c) $\binom{22}{11}$.

1.6.2. (a) **Lemma.** The exponent of a prime p in the canonical decomposition of $n!$ is equal to $\sum\limits_{i=1}^{\infty} \left[\frac{n}{p^i}\right]$.
(b) $n!$ is not divisible by 2^n for any $n \geq 1$.
(c) How many zeros are there at the end of 1000!?

1.6.3. Let $n = p_1^{\alpha_1} \cdot \ldots \cdot p_n^{\alpha_n}$ be the canonical decomposition. Find
(a) the number $\alpha(n)$ of all positive divisors of the number n;
(b) the sum $s(n)$ of all positive divisors of the number n;
(c) $\sum\limits_{d \mid n} \alpha(d)$, where the sum is taken over all positive divisors of the number n.

1.6.4. (a) Suppose that $(a, b) = 15$ and $[a, b] = 840$. Find a and b.
(b) Prove that $(a, b) \cdot [a, b] = ab$.
(c) Express $[a, b, c]$ in terms of a, b, c, (a, b), (b, c), (c, a), and (a, b, c).
(d) Express (a, b, c) in terms of a, b, c, $[a, b]$, $[b, c]$, $[c, a]$, and $[a, b, c]$.
(e)* Find expressions similar to the ones above for n integers.

1.6.5. A positive number is said to be *perfect* if it is equal to the sum of all of its positive divisors other than itself. Prove that n is an even perfect number if and only if $n = 2^{p-1}(2^p - 1)$, where p and $2^p - 1$ are primes.

1.6.6. (a) If $(a, b) = 1$ and $ab = m^2$, then there exist k and l such that $a = k^2$ and $b = l^2$.
(b) Find $n > m > 100$ such that $1 + 2 + \ldots + n = m^2$.
(c) Find all $m > n > 1$ such that $1^2 + 2^2 + \ldots + n^2 = m^2$.
(d) If $n > 2$, $ab = c^n$, and $(a, b) = 1$, then $a = x^n$ and $b = y^n$ for some x and y.
(e) The integer $m(m + 1)$ is not a power of a prime number for any $m > 1$.

1.6.7. (a) If $ab = cd$, then there exist k, l, m, and n such that $a = kl$, $b = mn$, $c = km$, and $d = ln$.
(b) Find all integers a, b, c, d, k, and m such that $ab = cd$, $a + d = 2^k$, and $b + c = 2^m$.

1.6.8. Find the smallest integer n such that for any set of n numbers between 1 and 200, there are a and b in the set with $a \mid b$.

1.6.9. (a) Let p be a prime and let $n < p < 2n$. Then $\binom{2n}{n}$ is divisible by p.

(b) The following inequality holds: $2^{2p_n+1} > p_1 \cdot \ldots \cdot p_n$, where p_n is the nth prime.

(c) **Bertrand's postulate.** For any $n > 1$ there exists a prime between n and $2n$.

Suggestions, solutions, and answers

1.6.1. (a) We have $1995 = 5 \cdot 399 = 5 \cdot 3 \cdot 133 = 5 \cdot 3 \cdot 7 \cdot 19 \; (= 5 \cdot 7 \cdot 57)$.

(b) Calculate the exponent of 2 in the canonical decomposition of $17! = 1 \cdot 2 \cdot 3 \cdot \ldots \cdot 17$. Every second number in this product is divisible by 2, so we can factor out 2^8. Then, each fourth number is divisible by 4, providing an additional factor of 2^4. Similarly we find two more 2's in factors of 8 and one more 2 in factors of 16. Applying this to the other primes yields

$$17! = 2^{15} \cdot 3^{\left[\frac{17}{3}\right]+\left[\frac{17}{9}\right]} \cdot 5^{\left[\frac{17}{5}\right]} \cdot 7^{\left[\frac{17}{7}\right]} \cdot 11 \cdot 13 \cdot 17 = 2^{15} \cdot 3^6 \cdot 5^3 \cdot 7^2 \cdot 11 \cdot 13 \cdot 17.$$

(c) Similarly to part (b) we have

$$11! = 2^{\left[\frac{11}{2}\right]+\left[\frac{11}{4}\right]+\left[\frac{11}{8}\right]} \cdot 3^{\left[\frac{11}{3}\right]+\left[\frac{11}{9}\right]} \cdot 5^{\left[\frac{11}{5}\right]} \cdot 7 \cdot 11 \quad \text{and}$$

$$22! = 2^{\left[\frac{22}{2}\right]+\left[\frac{22}{4}\right]+\left[\frac{22}{8}\right]+\left[\frac{22}{16}\right]} \cdot 3^{\left[\frac{22}{3}\right]+\left[\frac{22}{9}\right]} \cdot 5^{\left[\frac{22}{5}\right]} \cdot 7^{\left[\frac{22}{7}\right]} \cdot 11^{\left[\frac{22}{11}\right]} \cdot 13 \cdot 17 \cdot 19.$$

Therefore

$$\binom{22}{11} = \frac{22!}{11! \cdot 11!} = 2^{19-16} \cdot 3^{9-8} \cdot 5^{4-4} \cdot 7^{3-2} \cdot 13 \cdot 17 \cdot 19 = 2^3 \cdot 3 \cdot 7 \cdot 13 \cdot 17 \cdot 19.$$

1.6.3. Solve this problem for a prime n, then for $n = p^\alpha$, then for $n = p_1 p_2$, and finally for the general case.

1.6.4. *Hint.* Use the inclusion-exclusion principle and canonical decomposition.

(c) *Answer:*

$$[a, b, c] = \frac{a \cdot b \cdot c \cdot (a, b, c)}{(a, b) \cdot (b, c) \cdot (c, a)}.$$

1.6.9. A proof can be found in [**Tik94**]. Most of the technical details there are not needed if we just want to prove Bertrand's postulate, rather than Chebyshev's theorem. See also [**AZ04**].

7. Integer points under a line (2*)

The problems in this section investigate the sum

$$f_\alpha(n) = \sum_{k=1}^{n} [\alpha k],$$

which gives the number of integer points with positive y-coordinate and x-coordinate between 1 and n that lie under the line $y = \alpha x$, where α is a positive real number. An algorithm for rational α is developed in problems 1.7.3 (a, b, c), while problems 1.7.1 and 1.7.2 are useful as warm-ups.

1.7.1. (a) Find $f_{\sqrt{2}}(4)$.

(b) Do there exist numbers $\alpha \neq \beta$ such that $f_\alpha(n) = f_\beta(n)$ for any n?

1.7.2. Find $f_\alpha(n)$

(a) if α is an integer;   (b) if 2α is an integer;   (c) if 3α is an integer;

(d) if $\alpha = u/n$ for given integers u and n.

(e) Prove that $\lim\limits_{n\to\infty} \frac{f_\alpha(n)}{n^2}$ exists, and find it. (See the definition of limits in problem 6.4.2; skip this problem if you are unfamiliar with this concept.)

1.7.3. (a) Prove the equality $f_\alpha(n) = f_{\{\alpha\}}(n) + \frac{1}{2}[\alpha]n(n+1)$ for arbitrary α and n.

(b) Prove the equality $f_\alpha(n) + f_{1/\alpha}([n\alpha]) - [n/q] = n[n\alpha]$, where q is the denominator of the irreducible fraction representing α if α is rational, and $q = \infty$ (i.e., $[n/q] = 0$) if α is irrational.

(c) Construct an algorithm for calculating $f_\alpha(n)$ for rational α, using (a) and (b).

(d) Find the complexity of that algorithm, that is, the number of operations of addition and multiplication in the algorithm, and compare it with the complexity of the straightforward calculation of $f_\alpha(n)$.

(e) Find an algorithm for calculating the sum $\sum\limits_{k=1}^{n} \{\alpha k\}$ for a rational α.

Remark 1.7.4. The special case of the equality in 1.7.3 (b) for odd positive relatively prime numbers $p < q$, $\alpha = p/q$, and $n = (q-1)/2$ (then $[n\alpha] = (p-1)/2$) appears in the proof of the quadratic reciprocity law (see the solution of problem 2.4.5 (d)). The proof in the general case is similar.

The sum from 1.7.3 (e) was calculated (in a more cumbersome way than proposed here) in [**Dob04**].

Suggestions, solutions, and answers

1.7.2. (a) We have $\sum\limits_{k=1}^{n} [\alpha k] = \alpha \sum\limits_{k=1}^{n} k = \alpha \cdot \frac{n(n+1)}{2}$.

(b) For integer α see (a). For half-integers ($\alpha = q/2$ where q is odd) we have

$$[\alpha] + [2\alpha] + [3\alpha] + \ldots + [n\alpha]$$

$$= \left(\alpha - \frac{1}{2}\right) + 2\alpha + \left(3\alpha - \frac{1}{2}\right) + \ldots = \alpha \cdot \frac{n(n+1)}{2} - \left[\frac{n+1}{2}\right].$$

There are other ways to write this sum, for example

$$[\alpha]\frac{n(n+1)}{2} + \{\alpha\}\frac{n^2 + (-1)^n}{2}.$$

(c) For integer α see (a). If α is not an integer we have

$$f_\alpha(n) = \begin{cases} \alpha \cdot \frac{n(n+1)}{2} - \left[\frac{n+1}{3}\right], & n \neq 3k+1; \\ \alpha \cdot \frac{n(n+1)}{2} - \left[\frac{n}{3}\right] - \{\alpha\}, & n = 3k+1. \end{cases}$$

Hint. If α is not an integer, we have

$$[\alpha] + [2\alpha] + [3\alpha] = \alpha + 2\alpha + 3\alpha - \frac{1}{3} - \frac{2}{3} = (1+2+3)\alpha - 1.$$

Solutions to (a), (b), (c), and (d) can be obtained using Pick's formula. See [**Sop**].

(e) *Answer:* $\alpha/2$.

1.7.3. (a) We have

$$f_\alpha(n) = \sum_{k=1}^n [\alpha k] = \sum_{k=1}^n [([\alpha] + \{\alpha\}) \cdot k] = \sum_{k=1}^n [[\alpha]k + \{\alpha\}k]$$

$$= \sum_{k=1}^n ([\alpha]k + [\{\alpha\}k]) = \sum_{k=1}^n [\alpha]k + \sum_{k=1}^n [\{\alpha\}k]$$

$$= [\alpha] \sum_{k=1}^n k + f_{\{\alpha\}}(n) = [\alpha]\frac{n(n+1)}{2} + f_{\{\alpha\}}(n).$$

(b) Calculate the number of integer points in the rectangular region $1 \leq x \leq n$, $1 \leq y \leq [n\alpha]$. The details are similar to the solution of problem 2.4.5 (d).

(c) For example,

$$f_{2/3}(n) = n\left[\frac{2n}{3}\right] + \left[\frac{n}{3}\right] - f_{3/2}\left(\left[\frac{2n}{3}\right]\right);$$

$$f_{3/2}\left(\left[\frac{2n}{3}\right]\right) = \frac{1}{2}\left[\frac{2n}{3}\right]\left(\left[\frac{2n}{3}\right] + 1\right) + f_{1/2}\left(\left[\frac{2n}{3}\right]\right);$$

$$f_{1/2}\left(\left[\frac{2n}{3}\right]\right) = \left[\frac{2n}{3}\right]\left[\frac{n}{3}\right] + \left[\frac{n}{3}\right] - f_2\left(\left[\frac{n}{3}\right]\right),$$

since $[[x]/n] = [x/n]$ for an integer $n > 0$ and so $\left[\frac{\left[\frac{2n}{3}\right]}{2}\right] = \left[\frac{n}{3}\right]$;

$$f_2\left(\left[\frac{n}{3}\right]\right) = \left[\frac{n}{3}\right]\left(\left[\frac{n}{3}\right] + 1\right).$$

Chapter 2

Multiplication modulo p

The results in this chapter used in the rest of the book are the Euler–Fermat Theorem (problems 2.1.1 and 2.1.5) and the Primitive Root Theorem (problem 2.5.6 (b)). However, to use the Primitive Root Theorem it is not necessary to understand its proof.

In this chapter all variables are integers or residues modulo a prime (the exact meaning of the term will be clear from the context).

1. Fermat's Little Theorem (2)

2.1.1. (a) Let $\mathbb{Z}_{97} = \{0, 1, \ldots, 96\}$. Define the mapping $f\colon \mathbb{Z}_{97} \to \mathbb{Z}_{97}$ as follows: $f(a)$ is the remainder on division of the number $14a$ by 97. Then f is a one-to-one correspondence.

Discussion. It is sufficient to prove either surjectivity or injectivity. Usually one proves *injectivity*. This proof is usually based on Lemma 1.5.7 (b), whose proof in turn stems from the solvability of the equation $97x + 14y = 1$, which immediately implies *surjectivity*.

(b) The following congruence holds: $(14 \cdot 1) \cdot (14 \cdot 2) \cdot \ldots \cdot (14 \cdot 96) \equiv 96!$ (mod 97).

(c) The following congruence holds: $14^{96} \equiv 1$ (mod 97).

(d) **Fermat's Little Theorem.** If p is prime, then $n^p - n$ is divisible by p for any integer n.

Alternative formulation. If p is a prime and n is not divisible by p, then $n^p - 1$ is divisible by p.

(e) For prime p, $\binom{p}{k}$ is divisible by p for all $k = 1, 2, \ldots, p-1$. (This can be used for another proof [by induction] of Fermat's Little Theorem.)

2.1.2. Find the remainder upon division of
(a) 2^{100} by 101;   (b) 3^{102} by 101;   (c) 8^{900} by 29;

(d) 3^{2000} by 43;   (e) 7^{60} by 143;   (f) $2^{60} + 6^{50}$ by 143.

2.1.3. (a) If p is a prime and $p > 2$, then $7^p - 5^p - 2$ is divisible by $6p$.

(b) The number $1 \ldots 1$ consisting of 2002 ones is divisible by 2003.

(c) If p and q are different primes, then $p^q + q^p - p - q$ is divisible by pq.

(d) The number $30^{239} + 239^{30}$ is composite.

(e) If p is a prime, then the length of the period of the decimal expansion of the fraction $1/p$ divides $p - 1$.

2.1.4. For a prime p and an integer (or a residue) a not divisible by p, the smallest number $k > 0$ such that $a^k \equiv 1 \pmod{p}$ is called the *order* of a modulo p and is denoted by $\operatorname{ord} a = \operatorname{ord}_p a$. In other words,

$$\operatorname{ord} a = \operatorname{ord}_p a := \min\{k \geq 1 \mid a^k \equiv 1 \pmod{p}\}.$$

(a) The set $\{m \geq 0 \mid a^m \equiv 1 \pmod{p}\}$ consists of non-negative multiples of $\operatorname{ord} a$.

(b) If $a^m \equiv a^n \pmod{p}$, then $m - n$ is divisible by $\operatorname{ord} a$.

(c) **Lemma.** The number $p - 1$ is divisible by $\operatorname{ord} a$.

(d) If $\operatorname{ord} x$ and $\operatorname{ord} y$ are relatively prime, then $\operatorname{ord}(xy) = \operatorname{ord} x \cdot \operatorname{ord} y$.

(e) Let a and x be any integers, and let p be any prime number. Is it true that $a \operatorname{ord}_p x^a = \operatorname{ord}_p x$?

Notice that we can define division and negative powers modulo a prime. Statements analogous to 2.1.4 (a, b) hold for negative powers.

2.1.5. In these problems, $p, q, p_1, \ldots, p_k$ denote different prime numbers.

(a) If $p \neq q$ and n is divisible neither by p nor by q, then $n^{(p-1)(q-1)} - 1$ is divisible by pq.

(b) If n is not divisible by p, then $n^{p^\alpha(p-1)} - 1$ is divisible by $p^{\alpha+1}$.

(c) **Euler's Theorem**. If n is relatively prime to $m = p_1^{\alpha_1} \cdot \ldots \cdot p_k^{\alpha_k}$ and $\varphi(m) := (p_1 - 1)p_1^{\alpha_1 - 1} \cdot \ldots \cdot (p_k - 1)p_k^{\alpha_k - 1}$, then $n^{\varphi(m)} - 1$ is divisible by m.

(d) The number $\varphi(m)$ is equal to the number of integers between 1 and m that are relatively prime to m.

2.1.6. (Challenge.) Let n be an odd integer between 3 and 47 that is not divisible by 5. How can we quickly calculate the unknown n if we know $n^7 \bmod 50$?

The solution of this challenge shows why cryptography requires efficient ways to find the prime decomposition of a number or to recognize if a number is prime.

Suggestions, solutions, and answers

2.1.1. (a) $14 \cdot 7k \equiv k \pmod{97}$.

(b) $(14 \cdot 1) \cdot (14 \cdot 2) \cdot \ldots \cdot (14 \cdot 96) \equiv f(1) \cdot f(2) \cdot \ldots \cdot f(96) = 96! \bmod 97$.

(c) Cancel out 96! from the equality in (b) .

2.1.2. *Answers:* (a) 1; (b) 9; (c) 7; (d) 15; (e) 1; (f) 24.

2.1.6. $n \equiv (n^7)^3 \bmod 50$.

2. Primality tests (3*)
By S. V. Konyagin

2.2.1. If $2^m - 1$ is prime, then the number m is prime.

2.2.2. (a) If $2^{2^n} + 1$ is divisible by d, then $d - 1$ is divisible by 2^{n+1}.

(b)* Using the equality $641 = 5^4 + 2^4 = 1 + 5 \cdot 2^7$, prove that $2^{2^5} + 1$ is composite.

2.2.3. (a) Let $p > 2$ be prime. If $2^p - 1$ is divisible by d, then $d - 1$ is divisible by $2p$. In other words, any divisor of the number $2^p - 1$ has the form $2kp + 1$.

(b) If $p > 2$ is prime and a is not divisible by p, then $a^{(p-1)/2} \equiv \pm 1 \bmod p$.

(c) If $n - 1$ is divisible by 2^s and $a^{(n-1)/2} + 1$ is divisible by n for some a, then any prime divisor of n has the form $2^s k + 1$.

(d) If $n = 2^s k + 1$, $k \leq 2^s$, and $a^{(n-1)/2} \equiv -1 \bmod n$ for some a, then n is prime.

We obtained a sufficient condition for primality for numbers of a special type. Notice that if the number $n = 2^s k + 1$ is actually prime, then as a rule it is possible to find a number a satisfying the congruence $a^{(n-1)/2} \equiv -1 \bmod n$ by a small search.

2.2.4. *Fermat's Little Theorem is not a sufficient condition for primality.*

(a) If $p \geq 5$ is prime, then $n = (2^{2p} - 1)/3$ is composite, but $2^{n-1} \equiv 1 \bmod n$.

(b)* Find at least one composite number n such that for any integer a, the equality $(a, n) = 1$ implies that $a^{n-1} \equiv 1 \bmod n$.

2.2.5. The Lucas test. A number $n = 2^m - 1 > 3$ is a prime if and only if $m > 2$ is a prime and M_{m-1} is divisible by n. Here *the Lucas sequence* is defined by the formulas $M_1 = 4$ and $M_k = M_{k-1}^2 - 2$.

The proof is outlined in the following problem. Before trying to solve it you may find it useful to solve problems 2.4.1–2.4.4 with the help of hints from S. V. Konyagin.

2.2.6. Let $p \geq 5$ be a prime number. Define $x_k^{\pm} = (2 + \sqrt{3})^k \pm (2 - \sqrt{3})^k$,

$$X^+ = \{k : x_k^+ \equiv 0 \bmod p\}, \quad \text{and} \quad X^- = \{k : x_k^-/\sqrt{3} \equiv 0 \bmod p\}.$$

(a) For any integer k, x_k^+ and $x_k^-/\sqrt{3}$ are integers.

(b) If $z_1, z_2 \in X^+$, then $z_1 + z_2 \in X^-$.

(c) If $z_1, z_2 \in X^-$, then $z_1 + z_2 \in X^-$.

(d) If $z_1 \in X^+$ and $z_2 \in X^-$, then $z_1 + z_2 \in X^+$.

(e) Either $p + 1 \in X^-$ or $p - 1 \in X^-$.

(f) If $X^+ \neq \emptyset$ and z is the smallest positive element of the set X^+, then $X^+ = \{(2k+1)z\}$ where k runs through the set of integers and $z < p$.

(g) If k is a prime, then $M_k = x_{2^k-1}^+$.

Hints

2.2.4. (b) Consider integers of the form $n = pqr$, where p, q, and r are distinct primes.

Suggestions, solutions, and answers

2.2.2. (a) If $d_1 \equiv d_2 \equiv 1 \bmod 2^{n+1}$ then $d_1 d_2 \equiv 1 \bmod 2^{n+1}$. Therefore we can assume that d is prime. The number $2^{2^{n+1}} - 1$ is divisible by d and $2^{2^n} - 1$ is not. Therefore $\operatorname{ord}_d 2$ (see the definition of ord_p in problem 2.1.4) divides 2^{n+1} and does not divide 2^n. Thus, $\operatorname{ord}_d 2 = 2^{n+1}$, and $\operatorname{ord}_d 2$ divides $d - 1$.

2.2.3. (b) Note that

$$a^{p-1} - 1 = \left(a^{\frac{p-1}{2}}\right)^2 - 1 = \left(a^{\frac{p-1}{2}} - 1\right)\left(a^{\frac{p-1}{2}} + 1\right)$$

is divisible by p by Fermat's Little Theorem. Therefore, one of the two factors $a^{\frac{p-1}{2}} - 1$ and $a^{\frac{p-1}{2}} + 1$ is divisible by p.

(c) Let p be a prime divisor of n. Let t be non-negative and let l be an odd number such that $\operatorname{ord}_p a = 2^t l$ (see the definition in problem 2.1.4). From 2.1.4 (c), $p - 1$ is divisible by $\operatorname{ord}_p a = 2^t l$. Therefore it is sufficient to show that $t \geq s$.

According to the statement of problem 2.1.4 (c), $n - 1 = 2^s k$ is divisible by $\operatorname{ord}_p a = 2^t l$. Therefore, l divides k. If $t < s$ then $2^t l$ divides $(n-1)/2 = 2^{s-1} k$. Therefore $a^{\frac{n-1}{2}} \equiv 1 \bmod p$, contradicting the fact that n divides $a^{\frac{n-1}{2}} + 1$.

(d) If n is composite, then it has a prime divisor $p \leq \sqrt{n}$. From (c) it follows that $p \geq 2^s + 1$, and thus $n \geq (2^s + 1)^2$. This contradicts the condition $n = 2^s k + 1 \leq (2^s)^2 + 1$.

2.2.4. (a) Clearly $2^p = 2 \cdot (2^2)^{\frac{p-1}{2}} \equiv 2 \bmod 3$. Therefore $n = \frac{(2^p+1)(2^p-1)}{3}$ is composite. Since $2^{2p} = 2^2 \cdot (2^{p-1})^2 \equiv 4 \bmod p$, we see that $2^{2p} - 4$ is divisible by $2p$. Since $p > 3$, $n - 1 = (2^{2p} - 4)/3$ is also divisible by $2p$. Consequently, $2^{n-1} = (2^{2p})^{\frac{n-1}{2p}} \equiv 1 \bmod (2^{2p} - 1)$. Thus, $2^{n-1} - 1$ is divisible by $2^{2p} - 1 = 3n$.

(b) *Answer*: For example, $n = 561$.

3. Quadratic residues (2*)

The goal of the problems in this section is to motivate and illuminate the problem of solvability of the congruence $x^2 \equiv a \pmod{p}$, where p is an odd prime.

2.3.1. (a) What are possible remainders when a perfect square is divided by

3, 4, 5, 6, 7, 8, 9, and 10?

(b) If $a^2 + b^2$ is divisible by 3 (by 7), then a and b are divisible by 3 (by 7).

(c) A number of the form $4k + 3$ is not representable as a sum of two squares.

(d) There are infinitely many numbers not representable as sums of three squares.

2.3.2. Solve the following equations in integers.

(a) $x_1^2 + x_2^2 + x_3^2 + x_4^2 + x_5^2 = y^2$ (in odd numbers);

(b) $3x = 5y^2 + 4y - 1$;

(c) $x^2 + y^2 = 3z^2$;

(d) $2^x + 1 = 3y^2$;

(e) $x^2 = 2003y - 1$;

(f) $x^2 + 1 = py$, where $p = 4k + 3$.

2.3.3. (a) If the prime $p = 4k + 3$ divides $a^2 + b^2$, then $p|a$ and $p|b$.

(b) If the canonical decomposition of a number contains a prime factor of the form $4k+3$ with an odd exponent, then this number cannot be expressed as the sum of two squares.

(c)* The equation $x^2 + 1 = py$ is solvable in integers if $p = 4k + 1$ (and not solvable if $p = 4k + 3$).

(d)* Any prime number of the form $4k + 1$ can be expressed as a sum of two squares.

(e)* If every prime factor of the form $4k + 3$ in the canonical decomposition of a number has an even exponent, then the number can be expressed as a sum of two squares.

(f) There are infinitely many primes of the form $4k + 1$.

A very short proof of part (d) was given by Don Zagier in [**Pra07a**].

2.3.4. (Challenge.) Reduce the equation $py = at^2 + bt + c$, $a \neq 0$, to the congruence $x^2 \equiv k \pmod{p}$.

A residue $a \neq 0$ is said to be a *quadratic residue* (*quadratic nonresidue*) *modulo p* if the congruence $x^2 \equiv a \pmod{p}$ is solvable (not solvable).

2.3.5. (a) Give an example of a and p such that a and $-a$ are both quadratic residues modulo p.

(b) If a is not divisible by p, then the congruence $x^2 \equiv a^2 \pmod{p}$ has exactly two solutions.

(c) **Lemma.** The number of quadratic residues is equal to the number of quadratic nonresidues and is equal to $\frac{p-1}{2}$.

2.3.6. (a) **Lemma.** For any $a \neq 0$ there exists a unique b such that $ab \equiv 1 \pmod{p}$.

Notation: $b = a^{-1}$.

(b) Solve the congruence $x \equiv x^{-1} \pmod{p}$.

(c) **Wilson's Theorem.** The number $(p-1)! + 1$ is divisible by p.

2.3.7. (a) If $a \neq 0$ is a quadratic residue, then a^{-1} is also a quadratic residue.

(b) The number of quadratic residues is even if and only if -1 is a quadratic residue.

Lemma 2.3.8. (a) The product of two quadratic residues is a quadratic residue.

(b) The product of a quadratic residue and a quadratic nonresidue is a quadratic nonresidue.

(c) The product of two quadratic nonresidues is a quadratic residue.

Hints

2.3.3. (c) If you have difficulty, come back to this problem after you have studied this section.

(e) Use the statement (d) without proof.

Suggestions, solutions, and answers

2.3.1. (a) *Answer*: The squares have the following remainders upon division

by 3: 0, 1;   by 4: 0, 1;   by 5: 0, 1, 4;

by 6: 0, 1, 3, 4;   by 7: 0, 1, 2, 4;   by 8: 0, 1, 4;

by 9: 0, 1, 4, 7;   by 10: 0, 1, 4, 5, 6, 9.

Solution. It is sufficient to find squares of the remainders. Notice that 0 and 1 are squares modulo any number. Also, note that k^2 and $(-k)^2$

have the same remainder on division by n, so we need only consider k^2 for $2 \leq k \leq n/2$. We have

$2^2 \equiv 0 \bmod 4;$ $2^2 \equiv 4 \bmod 5;$

$2^2 \equiv 4, \ 3^2 \equiv 3 \bmod 6;$ $2^2 \equiv 4, \ 3^2 \equiv 2 \bmod 7;$

$2^2 \equiv 4, \ 3^2 \equiv 1, \ 4^2 \equiv 0 \bmod 8;$ $2^2 \equiv 4, \ 3^2 \equiv 0, \ 4^2 \equiv 7 \bmod 9;$

$2^2 \equiv 4, \ 3^2 \equiv 9, \ 4^2 \equiv 6, \ 5^2 \equiv 5 \bmod 10.$

(b:3) Considering divisibility by 3, assume the opposite. Then, according to (a), the remainders upon division by 3 of a^2 and b^2 are both equal to 1. Therefore $a^2 + b^2$ is not divisible by 3.

(b:7) Considering divisibility by 7, assume the opposite. By (a), the remainder upon division by 7 of a^2 is equal to 1, 2, or 4. Then the remainder upon division by 7 of b^2 is equal to 6, 5, or 3 respectively. This contradicts (a).

(c) By (a), the remainder upon division by 4 of x^2 is equal to 0 or 1. Thus, the remainder upon division by 4 of the sum of two squares is equal to 0, 1, or 2.

2.3.2. (b) *Answer*: $\{(3k - 1, 15k^2 - 6k)\} = \{(3k + 2, 15k^2 + 24k + 9)\}$.

(e, f) Use Fermat's Little Theorem.

2.3.5. (c) The number of quadratic residues does not exceed $\frac{p-1}{2}$, because $a^2 \equiv (-a)^2 (p)$.

Suppose that there exist $1 \leq l < k \leq \frac{p-1}{2}$ such that $k^2 \equiv l^2 \pmod{p}$. Then one of the numbers $k - l$ and $k + l$ is divisible by p. But $0 < k - l < k + l < p$, a contradiction. Consequently, the number of residues is exactly equal to $\frac{p-1}{2}$, and the number of nonresidues is $p - 1 - \frac{p-1}{2} = \frac{p-1}{2}$.

2.3.8. (c) In contrast to (a) and (b) we do not employ a direct proof. Use (a), (b), and Lemma 2.3.5 (c).

4. The law of quadratic reciprocity (3*)

Here we build on the previous section to develop an algorithm for determining the solvability of the congruence $x^2 \equiv a \pmod{p}$ for a prime p.

2.4.1. If the number $p = 8k + 5$ is a prime, then
    (a) $2^{4k+2} \equiv -1 \pmod{p}$;
    (b) the equation $x^2 - 2 = py$ is not solvable in integers.

2.4.2. If the number $p = 8k + 1$ is a prime then
    (a) $2^{4k} \equiv 1 \pmod{p}$;
    (b) The equation $x^2 - 2 = py$ is solvable in integers.

2.4.3. (a) If the number $p = 8k \pm 1$ is a prime, then $2^{(p-1)/2} \equiv 1 \pmod{p}$.
(b) If the number $p = 8k \pm 3$ is a prime, then $2^{(p-1)/2} \equiv -1 \pmod{p}$.
(c) For which primes p is the equation $x^2 - 2 = py$ solvable in integers?

2.4.4. (a) If the number $p = 12k \pm 1$ is a prime, then $3^{(p-1)/2} \equiv 1 \pmod{p}$.
(b) If the number $p = 12k \pm 5$ is a prime, then $3^{(p-1)/2} \equiv -1 \pmod{p}$.
(c) For which primes p is $x^2 - 3 = py$ solvable in integers?

2.4.5. For each residue a and odd prime p we define the *Legendre symbol*

$$\left(\frac{a}{p}\right) := \begin{cases} +1 & \text{if } a \text{ is a quadratic residue modulo } p; \\ -1 & \text{if } a \text{ is a quadratic nonresidue modulo } p. \end{cases}$$

For example, $\left(\frac{2}{p}\right) = (-1)^{(p^2-1)/8}$ by problem 2.4.3 and $\left(\frac{ab}{p}\right) = \left(\frac{a}{p}\right)\left(\frac{b}{p}\right)$ by problem 2.3.8.

(a) **Euler's criterion.** The following congruence holds:

$$\left(\frac{a}{p}\right) \equiv a^{\frac{p-1}{2}} \pmod{p}.$$

(b) **Gauss's lemma.** The following equation holds:

$$\left(\frac{a}{p}\right) = (-1)^{\sum\limits_{x=1}^{(p-1)/2} \left[\frac{2ax}{p}\right]}.$$

(c) For any odd number a, the following equation holds:

$$\left(\frac{a}{p}\right) = (-1)^{\sum\limits_{x=1}^{(p-1)/2} \left[\frac{ax}{p}\right]}.$$

(d) **The law of quadratic reciprocity.** If p and q are odd primes, then

$$\left(\frac{q}{p}\right) = (-1)^{\frac{p-1}{2}\cdot\frac{q-1}{2}} \left(\frac{p}{q}\right).$$

(e) Devise an algorithm for calculating $\left(\frac{a}{p}\right)$ and estimate its complexity (complexity is defined in problem 1.7.3 (d)).

2.4.6. If p is a prime and n and a are integers with $n > 0$, then the congruence $x^n \equiv a \pmod{p}$ has no more than n solutions. (If you cannot solve this problem, see problem 3.3.5 (f).)

Suggestions, solutions, and answers

Solutions of problems 2.4.1 (a), 2.4.2 (a), and 2.4.5 are based on K. Oganesyan's texts.

2.4.1. (a) Let $X := \{1, 2, \ldots, 4k+2\} - \{2 \cdot 1, 2 \cdot 2, \ldots, 2(2k+1)\}$. Then the sets of remainders upon division by $p = 8k + 5$ in the sets $\{2(2k+2), 2(2k+3), \ldots, 2(4k+2)\}$ and $-X$ coincide. Therefore $2^{4k+2} \cdot (4k+2)! \equiv -(4k+2)!$ (mod p).

In other words,

$$(8k + 4)! \equiv 2^{4k+2} \cdot 1 \cdot 2 \cdot \ldots \cdot (4k+2) \cdot 1 \cdot 3 \cdot \ldots$$
$$\cdot (8k + 3)2^{4k+2}(4k+2)! \cdot (-1)^{2k+1}(8k+5-1)(8k+5-3) \cdot \ldots$$
$$\cdot (8k + 5 - (4k+1)) \cdot (4k+3)(4k+5) \cdot \ldots \cdot (8k+3)$$
$$\equiv 2^{4k+2}(-1)(8k+4)! \pmod{p}.$$

Note: Solutions to problems 2.4.2(a), 2.4.3(a, b), 2.4.4(a, b), 2.4.5(b) are similar to the solution of problem 2.4.1(a).

2.4.2. (a) We have

$$-(8k)! \equiv -2^{4k} \cdot 1 \cdot 2 \cdot \ldots \cdot 4k \cdot 1 \cdot 3 \cdot \ldots$$
$$\cdot (8k - 1)(-1)2^{4k}(4k)! \cdot (-1)^{2k}(8k+1-1)(8k+1-3) \cdot \ldots$$
$$\cdot (8k + 1 - (4k-1)) \cdot (4k+1)(4k+3) \cdot \ldots \cdot (8k-1)$$
$$\equiv -2^{4k}(-1)^{2k}(8k)! \pmod{p}.$$

2.4.3. (c) *Answer*: $p = 8k \pm 1$.

2.4.4. (c) *Answer*: $p = 12k \pm 1$.

2.4.1 (a), **2.4.2** (a), **2.4.3** (a, b). *Hint (by S. V. Konyagin).* Let $z = (1 + i)\sqrt{2}/2$. Then $(z + 1/z)^p - (z^p + 1/z^p)$ can be expressed in the form $p(A + B\sqrt{2})$, where A and B are integers.

2.4.4. (a, b) *Hint (by S. V. Konyagin).* Let $z = (1 + i\sqrt{3})/2$. Then $(z + 1/z)^p - (z^p + 1/z^p)$ can be written in the form $p(A + B\sqrt{3})$, where A and B are integers.

2.4.5. (a) Denote by R (respectively, Q) the product of all residues (respectively, nonresidues) modulo p. If a is a quadratic nonresidue, then $a^{\frac{p-1}{2}} R \equiv Q$ (mod p). From problem 2.3.7 (a) we have $R \equiv \pm 1$ (mod p). Then Wilson's Theorem yields $Q \equiv -R$ (mod p). Thus, $a^{\frac{p-1}{2}} \equiv -1$ (mod p).

(a) *Another solution.* Suppose, to the contrary, that a is a quadratic nonresidue, and $a^{\frac{p-1}{2}} \equiv 1$ (mod p). Then the polynomial $x^{\frac{p-1}{2}} - 1$ over $\mathbb{Z}_p$ has more than $\frac{p-1}{2}$ roots, contradicting 2.4.6.[1]

(c) Use the equalities $\left(\frac{a}{p}\right) = \left(\frac{2}{p}\right)\left(\frac{\frac{a+p}{2}}{p}\right)$ and $\left(\frac{2}{p}\right) = (-1)^{\frac{p^2-1}{8}}$.

---

[1]See p. xx for a definition of $\mathbb{Z}_p$.

(d) From (c) we have $\left(\frac{p}{q}\right)\left(\frac{q}{p}\right) = (-1)^{\sum_{x=1}^{(q-1)/2} \left[\frac{px}{q}\right] + \sum_{y=1}^{(q-1)/2} \left[\frac{py}{p}\right]}$. It is sufficient to show that the following equality holds:

$$\sum_{x=1}^{(q-1)/2} \left[\frac{px}{q}\right] + \sum_{y=1}^{(q-1)/2} \left[\frac{py}{p}\right] = \frac{(p-1)(q-1)}{4}.$$

To prove it, consider the rectangle $1 \le x \le \frac{p-1}{2}$, $1 \le y \le \frac{q-1}{2}$. On the straight line $y = qx/p$ there are no integer points. Since the number of integer points over the given line with the ordinate y is equal to $[py/q]$, the total number of integer points over the line inside of the rectangle is equal to $\sum_{x=1}^{(q-1)/2} \left[\frac{px}{q}\right]$. Similarly, the number of integer points under the given line inside the rectangle is equal to $\sum_{y=1}^{(q-1)/2} \left[\frac{py}{p}\right]$. The total number of integer points inside the considered rectangle is $(p-1)/2 \cdot (q-1)/2$.

(e) Use (a, b, c, d) above.

5. Primitive roots (3*)

2.5.1. Let a and b be relatively prime to m. Formulate and justify an algorithm for solving the congruence $a^x \equiv b \pmod{m}$ for $m \in \{2,3,4,5,6,7\}$. *(Analysis of similar congruences is one of the main motivations for this section.)*

2.5.2. (a) If $(a, 35) = 1$, then $a^{12} \equiv 1 \pmod{35}$.

(b) If m is divisible by two different odd prime numbers and $(a, m) = 1$, then $a^{\frac{\varphi(m)}{2}} \equiv 1 \pmod{m}$.[2]

Let $(g, m) = 1$. A residue g is said to be a *primitive root* modulo m if $g^1, g^2, \ldots, g^{\varphi(m)} \equiv 1$ are distinct $\pmod{m}$. For example,
- 2 is a primitive root modulo 5 but 4 is not;
- from 2.5.2 (b) we see that if m is divisible by two different odd prime numbers, then there does not exist a primitive root modulo m.

2.5.3. Prove the existence of a primitive root modulo a prime for primes of the following forms:

(a) 257;   (b) $2^l + 1$;   (c) $2^k \cdot 3^l + 1$;   (d) 151;   (e) $2^k \cdot 3^l \cdot 5^m + 1$.

---

[2]Recall that $\varphi(m)$ is defined to be the number of positive integers less than or equal to m that are relatively prime to m (see p. 18).

There is a simple way of solving (a), (b), and (c) that does not extend to (d) and (e). We will show how to solve (d) and (e) by examples.

2.5.4. (a) The residue g is a primitive root modulo 97 if and only if neither g^3 nor g^{32} is congruent to 1 modulo 97.

(b) The congruence $x^3 \equiv 1 \pmod{97}$ has exactly 3 solutions.

(c) The congruence $x^{32} \equiv 1 \pmod{97}$ has exactly 32 solutions.

(d) There exists a primitive root modulo 97.

(e) The number of primitive roots modulo 97 is equal to 63.

2.5.5. (a) The residue g is a primitive root modulo 151 if and only if neither g^{30} nor g^{50} nor g^{75} is congruent to 1 modulo 151.

(b) The congruence $x^k \equiv 1 \pmod{151}$ has exactly k solutions for $k \in \{30, 50, 75\}$.

(c) The following equivalence holds:

$$\begin{cases} x^{30} \equiv 1 \pmod{151} \\ x^{50} \equiv 1 \pmod{151} \end{cases} \iff x^{10} \equiv 1 \pmod{151}.$$

(d) There exists a primitive root modulo 151.

(e) The number of primitive roots modulo 151 is equal to 40.

2.5.6. (a) If p is a prime and $p - 1$ is divisible by d, then the congruence $x^d \equiv 1 \pmod{p}$ has exactly d solutions.

(b) **Primitive Root Theorem.** For any prime p there exists a number g such that the residues modulo p of $g^1, g^2, g^3, \ldots, g^{p-1} = 1$ are distinct.

(c) How many primitive roots are there modulo a prime p?

Suggestions, solutions, and answers

2.5.3. (b) If there is no primitive root, then the congruence $x^{2^{l-1}} \equiv 1 \pmod{p}$ has $p - 1 = 2^l > 2^{l-1}$ solutions.

2.5.6. (a) Notice that the polynomial $x^{p-1} - 1$ over $\mathbb{Z}_p$ has exactly $p - 1$ roots and is divisible by $x^d - 1$. Prove that if a polynomial of degree a has exactly a roots and is divisible by a polynomial of degree b, then the polynomial of degree b has exactly b roots.

Another solution may by obtained by noticing that if $p = kd$ then for any a the congruence $y^k \equiv a \pmod{p}$ has no more than k solutions.

(c) *Answer:* $\varphi(p - 1)$.

6. Higher degrees (3*)
By A. Ya. Kanel-Belov and
A. B. Skopenkov

2.6.1. (a) For each integer n and each odd k, the number $k^{2^n} - 1$ is divisible by 2^{n+2}.

(b) For any integer n, $2^{3 \cdot 7^n} - 1$ is divisible by 7^{n+1}.

2.6.2. For which numbers a it is true that
    (a) $2^a - 1$ is divisible by 3^{100};   (b) $2^a + 1$ is divisible by 3^{100};
    (c) $5^a - 1$ is divisible by 2^{100};   (d) $2^a - 1$ is divisible by 5^{100}?

Statement 2.6.1 (a) means that for any $n \geq 3$ there are no primitive roots modulo 2^n (see the definition in section 5). The answers to problem 2.6.2(a, d, c) and statement 2.6.1(b) mean that for any number n, 2 is a primitive root modulo 3^n and modulo 5^n, but 5 and 2 are not primitive roots modulo 2^n and modulo 7^n.

2.6.3. (a) Find a primitive root modulo 7^{100}.

(b) **Theorem.** Primitive roots exist only for moduli 2, 4, p^n, and $2p^n$.

2.6.4. Let $p > 2$ be prime and g a primitive root modulo p, and suppose that $g^{p-1} - 1$ is not divisible by p^2. Then g is a primitive root modulo
    (a) p^2;   (b) p^3;   (c) p^n for any n.

2.6.5. Let $p > 2$ be prime.

(a) If g is a primitive root modulo p, then one of $g^{p-1} - 1$ or $(g+p)^{p-1} - 1$ is not divisible by p^2.

(b) If g is a primitive root modulo p^2, then g is a primitive root modulo p^n for any n.

(c) For any positive integer n, there exists a primitive root modulo p^n.

(d) The same is true for modulo $2p^n$.

2.6.6. Lemma about increasing the exponent. Let p be prime, with $p > 2$ or $n > 1$, and let q not be divisible by p. Also, suppose that $x - 1$ is divisible by p^n but not by p^{n+1}. Then

(a) the number $x^q - 1$ is divisible by p^n, but not by p^{n+1};

(b) the number $x^p - 1$ is divisible by p^{n+1}, but not by p^{n+2};

(c) the number $x^{p^k q} - 1$ is divisible by p^{n+k}, but not by p^{n+k+1}. (A closely related statement is called *Hensel's lemma*.[3])

---

[3] Hensel's lemma allows one to "lift" a solution x of $f(x) \equiv 0 \pmod{p^{k-1}}$ to a new solution y of $f(y) \equiv 0 \pmod{p^k}$, where p is a prime and f a polynomial with integer

2.6.7. Find the length of the period of the decimal expansion of the fractions
(a) $1/3^{100}$;   (b) $1/7^{100}$.

2.6.8. (a) Each cyclic permutation of the digits in the period of the fraction $1/7 = 0.(142857)$ yields the fractions $1/7$, $2/7$, $3/7$, $4/7$, $5/7$, and $6/7$. Generalize this to all fractions $1/p$ with period length $p - 1$.
(b) Find all values of the remainders when 10^k is divided by 3^{100}.
(c) Prove that any combination of 20 consecutive digits can be found in the decimal expansion of $1/3^{100}$.

2.6.9. Find an integer n such that among the last 1000 digits of the number 2^n one can find 100 consecutive   (a) zeros;   (b)* nines.

2.6.10. Same questions as in 2.6.9 for 5^n.

Hints

2.6.2. (a) $2^2 = 1 + 3$.   (b) Use the result from part (a).   (c) $5 = 2^2 + 1$.

Suggestions, solutions, and answers

2.6.2. (a) *Answer:* For $2 \cdot 3^{99} | a$.

Hint. It's sufficient to prove that $2^{2 \cdot 3^k}$ *is the smallest nonzero degree of 2 which has remainder 1 when divided by* 3^{k+1}. This follows from the congruence $2^{2 \cdot 3^k} \equiv 3^{k+1} + 1 \pmod{3^{k+2}}$. We can prove this congruence by induction on k. The base case $k = 0$ is easily verified. Now suppose that the congruence is true for $k \geq 0$. Then $2^{2 \cdot 3^k} = t3^{k+1} + 1$, where $t \equiv 1 \pmod 3$. Therefore

$$2^{2 \cdot 3^{k+1}} = (2^{2 \cdot 3^k})^3 = t^3 3^{3k+3} + t^2 3^{2k+3} + t3^{k+2} + 1 \equiv 3^{k+2} + 1 \pmod{3^{k+3}}.$$

Hint for another solution. By induction on k prove that $\min\{a\colon 2^a \equiv 1 \pmod{3^k}\} = 2 \cdot 3^{k-1}$ and $2^{2 \cdot 3^{k-1}} - 1$ is not divisible by 3^{k+1}.
(c) *Answer:* For $2^{98} | a$.
Hint. It is enough to prove that 5^{2^k} *is the smallest nonzero degree of the number 5 that has remainder 1 upon division by* 2^{k+2}. This follows from the congruence $5^{2^k} \equiv 2^{k+2} + 1 \pmod{2^{k+3}}$. We can prove this congruence by induction on k. The base case $k = 0$ is easily verified. Suppose the

---

coefficients. More precisely, if p and $f'(x)$ are relatively prime, then $y = x + up^{k-1}$, where k satisfies $f(x)/p^{k-1} + uf'(x) \equiv 0 \pmod p$ and f' is the derivative of f (see p. 113).

congruence is true for $k \geq 0$. Then $5^{2^k} = t2^{k+2}+1$, where t is odd. Therefore
$$5^{2^{k+1}} = (5^{2^k})^2 = t^2 2^{2k+4} + t2^{k+3} + 1 \equiv 2^{k+3} + 1 \pmod{2^{k+4}}.$$

(d) *Answer*: For $4 \cdot 5^{99} | a$.

Chapter 3

Polynomials and complex numbers

Results in this chapter to be used later in the book are Bezout's Theorem and its applications (problems 3.3.4 (a, b) and 3.3.5), the trigonometric form of complex numbers (problem 3.5.4), and a few simple facts (e.g., problem 3.3.3).

In this section, "solve the equation or inequality" means "find *all real solutions*." Those not familiar with trigonometric functions may skip problems whose formulation involves such functions.

1. Rational and irrational numbers (1)

A number is called *rational* if it is a quotient of two integers, and otherwise is called *irrational*.[1]

3.1.1. Are the following numbers rational?
- (a) $\sqrt{2}$;
- (b) $\sqrt[n]{k}$, where the integer $k \geq 2$ is not the nth power of an integer;
- (c) $\sqrt{2} + \sqrt{3}$;
- (d) $\sqrt[7]{1 + \sqrt[3]{2} + \sqrt{3}}$;
- (e) $\frac{\sqrt{2} + \sqrt{3}}{\sqrt{2} - \sqrt{3}} + 2\sqrt{6}$;
- (f) $\sqrt{3 + 2\sqrt{2}} - \sqrt{2}$;
- (g) $\sqrt[3]{\sqrt{5} + 2} - \sqrt[3]{\sqrt{5} - 2}$;
- (h) $\sqrt{2} + \sqrt[3]{2}$;
- (i) $\sqrt{2} + \sqrt{3} + \sqrt{5}$;
- (j) $\sqrt{p_1} + \ldots + \sqrt{p_n}$, where $p_1, \ldots, p_n$ are different prime numbers.

3.1.2. The numbers $\sqrt{2}$, $\sqrt{3}$, and $\sqrt{5}$ are not members of any arithmetic progression (in any order; not even non-consecutive members).

Theorem 3.1.3. Let $A(x) = a_n x^n + a_{n-1} x^{n-1} + \ldots + a_1 x + a_0$ be a polynomial with integer coefficients.

---

[1] Recall that the set of rational numbers is denoted by $\mathbb{Q}$ (see p. xx).

(a) **On integer roots.** If $A(p) = 0$ for an integer $p \neq 0$, then p divides a_0.

(b) **On rational roots.** If $A(p/q) = 0$ for an irreducible fraction $p/q \neq 0$, then p divides a_0 and q divides a_n.

(c) If $A(p/q) = 0$ for an irreducible fraction p/q, then for any integer k the number $A(k)$ is divisible by $p - kq$.

3.1.4. Are the following numbers rational?
(a) $\cos 60°$; (b) $\sin 60°$; (c) $\cos 36°$; (d) $\cos 20°$;
(e) $\sin 10°$; (f) $\cos(2\pi/7)$; (g)* $\arccos(\frac{1}{3})/\pi$.

3.1.5. Prove the following equalities assuming that the angles α, 2α, 3α, β, $\alpha + \beta$, and $\alpha - \beta$ are acute.
(a) $\cos 2\alpha = 2\cos^2 \alpha - 1$;
(b) $\sin 2\alpha = 2\sin \alpha \cos \alpha$;
(c) $\cos(\alpha + \beta) = \cos \alpha \cos \beta - \sin \alpha \sin \beta$;
(d) $\sin(\alpha + \beta) = \sin \alpha \cos \beta + \cos \alpha \sin \beta$;
(e) $\cos 3\alpha = 4\cos^3 \alpha - 3\cos \alpha$;
(f) $\sin 3\alpha = 3\sin \alpha - 4\sin^3 \alpha$;
(g) $\cos(\alpha + \beta) + \cos(\alpha - \beta) = 2\cos \alpha \cos \beta$.

3.1.6. (a) For any n there is a polynomial T_n with integer coefficients such that $T_n(\cos x) = \cos nx$ for any x.
(b) Find the constant term of the polynomial T_n.
(c) Find the leading term of the polynomial T_n.

3.1.7. For which integers m and n is each of the following numbers rational?
(a) $\cos(2\pi/n)$; (b) $\cos n°$; (c) $\cos(2\pi m/n)$.

Suggestions, solutions, and answers

3.1.1. *Answers*: (a, b, c, d, h) No; (e, f, g) Yes.

Hints. (a) Assume the opposite: suppose $\sqrt{2} = p/q$, where p/q is an irreducible fraction. Square the equation and multiply both sides by q^2, thus getting $2q^2 = p^2$. Since p and q are integers, we conclude that p is even. Thus, $p = 2r$, where r is an integer. Substitute this into our equality to get $2q^2 = (2r)^2$. Divide both sides of this by 2, yielding $q^2 = 2r^2$. Again, we conclude that q is even. This contradicts the fact that the fraction p/q is irreducible, which proves that $\sqrt{2}$ is irrational.

(e) We have $\frac{\sqrt{2}+\sqrt{3}}{\sqrt{2}-\sqrt{3}} = \frac{(\sqrt{2}+\sqrt{3})(\sqrt{2}+\sqrt{3})}{(\sqrt{2}-\sqrt{3})(\sqrt{2}+\sqrt{3})} = \frac{5+2\sqrt{6}}{-1} = -5 - 2\sqrt{6}$.

(f) We have $\sqrt{3 + 2\sqrt{2}} - \sqrt{2} = 1$.

(g) We have $\sqrt[3]{\sqrt{5}+2} - \sqrt[3]{\sqrt{5}-2} = 1$; see problem 3.2.2 (a).

(h) *First hint.* Raise both sides of the equality $r - \sqrt{2} = \sqrt[3]{2}$ to the third power and get a contradiction.

Second hint. The number $\sqrt{2} + \sqrt[3]{2}$ is a root of the polynomial $((x - \sqrt{2})^3 - 2)((x + \sqrt{2})^3 - 2)$ with integer coefficients. By the rational roots theorem 3.1.3 (b), this equation does not have rational roots.

(i, j) See instruction for problem 8.3.1 (f).

3.1.2. The number $\frac{\sqrt{5} - \sqrt{3}}{\sqrt{3} - \sqrt{2}}$ is irrational.

3.1.3. (a) In the given equality all terms except a_0 are divisible by p.

(c) See Bezout's Theorem 3.3.4.

3.1.4. *Answers*: (b, c, d, e, f, g) No; (a) Yes.

Hints.

(a) $\cos 60° = 1/2$.

(b) $\sin 60° = \sqrt{3}/2$.

(c) $\cos 36° = (\sqrt{5} + 1)/4$.

(d) Using formula 3.1.5 (e) for the cosine of the triple angle yields $1/2 = \cos(\pi/3) = 4\cos^3(\pi/9) - 3\cos(\pi/9)$. If for an irreducible fraction p/q the equality $4(p/q)^3 - 3(p/q) = 1/2$ holds, then $8p^3 - 6pq^2 - q^3 = 0$, so 1 is divisible by p and 8 is divisible by q (this is a special case of Theorem 3.1.3 (b) on rational roots). Therefore $p/q \notin (1/2, 1)$. But $\cos(\pi/9) \in (1/2, 1)$.

(e) The solution is similar to (d) or can be reduced to (d) using $\cos 20° = 1 - 2\sin^2 10°$.

(f) Similar to (d). Use the condition $\cos(2\pi/7) \in (1/2, 1)$ and problem 3.1.6 (a, b).

(g) Similar to (d), using problem 3.1.6 (a, c).

3.1.5. (a, b) Consider an isosceles triangle with vertex angle 2α.

3.1.6. (a) Induction on n using formula 3.1.5(g).

3.1.7. (a) For odd $n \neq 5$, the solution is similar to 3.1.4 (f). For even $n \neq 8$, the solution reduces to the $n/2$ case by applying the equality $\cos(2\pi/n) = 2\cos^2(\pi/n) - 1$.

Answer: $n \in \{1, 2, 3, 4, 6\}$.

(c) Suppose $\cos(2\pi m/n)$ is a rational number for an irreducible fraction m/n. Then there exists k such that $mk \equiv 1 \pmod{n}$. Consequently, $\cos(2\pi mk/n) = \cos(2\pi/n)$ is rational.

Hint for an alternative solution. If $\cos\alpha \in \mathbb{Q} - \{\pm 1/2, \pm 1\}$, then the denominator of the irreducible fraction representing $\cos 2\alpha$ is bigger than the denominator of $\cos\alpha$. On the other hand, there exist integers $a, b > 0$ for which $2^a - 2^b$ is divisible by n.

2. Solving polynomial equations of the third and fourth degrees (2)

The author thanks O. E. Orel for useful discussions.

The material presented here is important and widely known, yet is not included in the school or university curriculum. Our treatment contrasts with other sources in that instead of unmotivated changes of variables, we show that equations can be naturally reduced to those whose solutions are clearly seen.

For example, the equation $x^2 + 4x - 1 = 0$ can be reduced to $y^2 - 5 = 0$ by the substitution $y = x + 2$.

3.2.1. (a) The equation $x^3 + 3x^2 + 5x + 7 = 0$ can be reduced, by a change of variable, to $y^3 + py + q = 0$ for some p and q.

(b) The equation $ax^3 + bx^2 + cx + d = 0$ with $a \neq 0$ can be reduced by a suitable change of variable to the form $y^3 + py + q = 0$ for some p and q.

(c) We can reduce $ax^4 + bx^3 + cx^2 + dx + c = 0$ with $a \neq 0$ by a suitable change of variable to the form $y^4 + py^2 + qy + r = 0$ for some p, q, and r.

3.2.2. (a) Prove that $\sqrt[3]{\sqrt{5} + 2} - \sqrt[3]{\sqrt{5} - 2} = 1$.

(b) Find at least one root of $x^3 - 3\sqrt[3]{2}x + 3 = 0$.

Hint. **The del Ferro method.** Since
$$(u + v)^3 = u^3 + v^3 + 3uv(u + v),$$
$u + v$ is a root of the equation $x^3 - 3uvx - (u^3 + v^3) = 0$.

(c) Solve the equation $x^3 - 3\sqrt[3]{2}x + 3 = 0$.

3.2.3. (a) Factor $a^3 + b^3 + c^3 - 3abc$.

(b) Prove the inequality $a^2 + b^2 + c^2 \geq ab + bc + ca$. When is equality achieved?

(c) Prove the inequality $a^3 + b^3 + c^3 \geq 3abc$ for $a, b, c > 0$.

(d) Factor $a^3 + b^3 + c^3 - 3abc$ into linear factors with complex coefficients.

For problems 3.2.4–3.2.7 below, one should have basic knowledge of complex numbers; for example, it is enough to be able to solve problems 3.5.1 and 3.5.2. Otherwise, feel free to skip these four problems.

3.2.4. (a) State and prove theorems describing all real (all complex) roots of $x^2 + px + q = 0$.

(b) State and prove theorems describing all real (all complex) roots of the equation $x^3 + px + q = 0$ in the case where the del Ferro method works (see problem 3.2.2). Under what condition is this method applicable if we only take square roots of positive numbers?

(c) Construct an explicit (i.e., symbolic) algorithm for finding all real roots of $ax^3 + bx^2 + cx + d = 0$, where $a \neq 0$.

When solving some cubic equations by the del Ferro method, complex numbers unexpectedly arise exactly in the case where all roots of the original equation are real. Such equations could be solved by the following purely real method. (Interestingly, this method also leads to *transcendental methods* of solving equations [**PS97**].)

3.2.5. Vieta's method. (a) Solve $4x^3 - 3x = \frac{1}{2}$.
   (b) Solve $x^3 - 3x - 1 = 0$.
   (c) Use the cosine and inverse cosine functions to devise a general formula for the solution of $x^3 + px + q = 0$ by the method outlined in these problems. Under what conditions can $x^3 + px + q = 0$ be solved by this method?

3.2.6. Solve
   (a) $(x^2 + 2)^2 = 9(x - 1)^2$;   (b) $x^4 + 4x - 1 = 0$;
   (c) $x^4 + 2x^2 - 8x - 4 = 0$;   (d) $x^4 - 12x^2 - 24x - 14 = 0$.

Hint for part 3.2.6 (b). **Ferrari's method.** Find a, b, and c such that
$$x^4 + 4x - 1 = (x^2 + a)^2 - (bx + c)^2.$$
To do so we must find at least one a such that $(x^2 + a)^2 - (x^4 + 4x - 1)$ is a perfect square. This leads to computing the discriminant of a quadratic polynomial. The discriminant is a cubic polynomial in a, called the *resolvent cubic* of $x^4 + 4x - 1$.

3.2.7.* (a) State and prove a theorem describing all real roots of the equation $x^4 + px^2 + qx + s = 0$. Use the resolvent cubic.
   (b) Do the same for all complex roots.
   (c) All complex roots of $x^4 + px^2 + qx + s = 0$ can be given by the following formula:
$$\pm\sqrt{2\alpha_1 - p} \pm \sqrt{2\alpha_2 - p} \pm \sqrt{2\alpha_3 - p},$$
where α_1, α_2, and α_3 are all roots of resolvent cubic, the number of minuses is even, and the values of the square roots are selected such that their product is equal to $-q$.

Hints

The hints below use material from [**ABG+**].

3.2.2. (a) Take cubes and use the identity $(u - v)^3 = u^3 - v^3 - 3uv(u - v)$.

3.2.3. (a) When $a = -b - c$, the polynomial is equal to zero. Then divide $a^3 - 3abc + (b^3 + c^3)$ by $a + b + c$ using "long division."

3.2.4. (b) *Answer*: The del Ferro method is applicable if $D_{pq} := \left(\frac{p}{3}\right)^3 + \left(\frac{q}{2}\right)^2 \geq 0$.

Theorem 3.2.8. Let $p, q \in \mathbb{R}$. If $D_{pq} > 0$, then $x^3 + px + q = 0$ has one real root

$$\sqrt[3]{-\frac{q}{2} + \sqrt{D_{pq}}} - \sqrt[3]{\frac{q}{2} + \sqrt{D_{pq}}}.$$

If $D_{pq} = 0$, then the real roots of the equation $x^3 + px + q = 0$ are $-2\sqrt[3]{q/2}$ and $-\sqrt[3]{q/2}$ (they are distinct if $q \neq 0$).

3.2.5. (a) Use statement 3.1.5 (a).

3.2.7. Use Ferrari's method (see problem 3.2.6 (b)). Take care to analyze all cases.

It is also possible to solve the equation $x^4 + ax^3 + bx^2 + cx + d = 0$ by selecting α, A, and B so that

$$x^4 + ax^3 + bx^2 + cx + d = \left(x^2 + \frac{ax}{2} + \alpha\right)^2 - (Ax + B)^2.$$

Suggestions, solutions, and answers

3.2.1. Make the substitution $y := x + \frac{b}{3a}$ in parts (a) and (b), and $y := x + \frac{b}{4a}$ in part (c).

3.2.2. (a) Let $x = \sqrt[3]{2 + \sqrt{5}} - \sqrt[3]{\sqrt{5} - 2}$. Then $x^3 = 4 - 3x$. This equation has the root $x = 1$, and since $x^3 + 3x - 4$ is monotone, there are no other (real) roots.

Another solution follows from the equality $\sqrt[3]{\sqrt{5} \pm 2} = (\sqrt{5} \pm 1)/2$.

(b) We have $x^3 - 3\sqrt[3]{2}x + 3 = x^3 - 3bcx + (b^3 + c^3)$, where $b = 1$ and $c = \sqrt[3]{2}$.

Answer: $x = -1 - \sqrt[3]{2}$.

(c) From the solution to 3.2.3 (a) we see that $x^3 - 3\sqrt[3]{2}x + 3 = 0$ is equivalent to

$$(x + b + c)(x^2 + b^2 + c^2 - bc - bx - cx) = 0, \quad \text{where} \quad b = 1 \quad \text{and} \quad c = \sqrt[3]{2}.$$

From 3.2.3 (b) we see that since $b \neq c$, the second factor in the product is positive for all x. Therefore the original equation has the unique root $x = -b - c = -1 - \sqrt[3]{2}$.

Answer: $x = -1 - \sqrt[3]{2}$.

3.2.3. (a, d) *Answer*:

$$a^3 + b^3 + c^3 - 3abc = (a + b + c)(a^2 + b^2 + c^2 - ab - bc - ca)$$

$$= (a + b + c)(a + b\varepsilon + c\varepsilon^2)(a + b\varepsilon^2 + c\varepsilon) \quad \text{where } \varepsilon = \frac{-1 + i\sqrt{3}}{2}.$$

(b) $2(a^2 + b^2 + c^2 - ab - bc - ca) = (a - b)^2 + (b - c)^2 + (c - a)^2 \geq 0$.
Equality is achieved if and only if $a = b = c$.

3.2.4. (b) *Proof of Theorem* 3.2.8. Write

$$u := -\sqrt[3]{\frac{q}{2} + \sqrt{D_{pq}}} \quad \text{and} \quad v := \sqrt[3]{-\frac{q}{2} + \sqrt{D_{pq}}}.$$

We have $uv = -p/3$ and $u^3 + v^3 = -q$. Using the formula from the solution
to 3.2.3 (a) for $a = x$, $b = -u$, and $c = -v$, we see that $u + v$ is a root of
the polynomial $x^3 + px + q = x^3 - 3uvx - u^3 - v^3$. Using the solution to
3.2.3 (b), if $D_{pq} > 0$ there are no other roots, and if $D_{pq} = 0$ there is another
root, $u = v = -\sqrt[3]{q/2}$.

Theorem. Let $p, q \in \mathbb{C}$ and $pq \neq 0$. Let
- $\sqrt{D_{pq}}$ denote any of the two values of the square root of D_{pq};
- u denote any of the three values of the cube root of $-\frac{q}{2} - \sqrt{D_{pq}}$;
- $v := -\frac{p}{3u}$ ($p \neq 0$, so $(q/2)^2 \neq D_{pq}$, so $u^3 = -\frac{q}{2} - \sqrt{D_{pq}} \neq 0$).

Then all roots of the equation $x^3 + px + q = 0$ are $u + v$, $u\varepsilon_3 + v\varepsilon_3^2$, and
$u\varepsilon_3^2 + v\varepsilon_3$ (not necessarily different).

Proof. We have $uv = -p/3$ and $u^3 + v^3 = -q$. The theorem follows by the
formula used in the solution to problem 3.2.3 (d) for $a = x$, $b = -u$, and
$c = -v$. □

3.2.5. (a) Similarly to 3.1.5 (b) $\cos\frac{\pi}{9}$, $\cos\frac{7\pi}{9}$, and $\cos\frac{13\pi}{9}$ are the roots of
$4y^3 - 3y = \frac{1}{2}$. By 3.3.5 (b) there are no other roots.

Answer: $x \in \left\{ \cos\frac{\pi}{9}, \cos\frac{7\pi}{9}, \cos\frac{13\pi}{9} \right\}$.

(b) The substitution $y = 2x$ reduces this to (a).

Answer: $x \in \left\{ 2\cos\frac{\pi}{9}, 2\cos\frac{7\pi}{9}, 2\cos\frac{13\pi}{9} \right\}$.

3.2.6. *Answers:* (b) $\frac{-\sqrt{2} \pm \sqrt{4\sqrt{2} - 2}}{2}$; (c) $\frac{\sqrt{2} \pm \sqrt{8\sqrt{2} - 6}}{2}$; (d) $\sqrt{2} \pm (\sqrt[4]{2} + \sqrt[4]{8})$.

3.2.7. If $q = 0$, then we have a biquadratic (fourth-degree) equation which
is easy to solve. So assume that $q \neq 0$.

(a) **Theorem.** Let $p, q, s \in \mathbb{R}$, with $q \neq 0$. Then there exists $\alpha > p/2$
such that $q^2 = 4(2\alpha - p)(\alpha^2 - s)$. For any such α define $A := \sqrt{2\alpha - p}$. All

real roots of the equation $x^4 + px^2 + qx + s = 0$ are described as follows:

$$\begin{cases} \text{no roots}, & 2\alpha + p > 2|q|/A; \\ x_{\pm}, \text{ where } x := \left(-A \pm \sqrt{-2\alpha - p + \frac{2q}{A}}\right)/2, & -2q/A < 2\alpha + p \le 2q/A; \\ y_{\pm}, \text{ where } y := \left(A \pm \sqrt{-2\alpha - p - \frac{2q}{A}}\right)/2, & 2q/A < 2\alpha + p \le -2q/A; \\ x_{\pm}, \ y_{\pm}, & 2\alpha + p \le -2|q|/A. \end{cases}$$

Proof. Let $R(x) := 4(2x - p)(x^2 - s) - q^2$. Then $R(p/2) = -q^2 < 0$. For sufficiently large x we have $R(x) > 0$. The Intermediate Value Theorem 7.1.13 implies that there exists $\alpha > p/2$ such that $R(\alpha) = 0$.

Since $p = 2\alpha - A^2$ and α are roots of the resolvent, we have $s = \alpha^2 - \frac{q^2}{4(2\alpha - p)} = \alpha^2 - \frac{q^2}{4A^2}$. Then

$$x^4 + px^2 + qx + s = \left(x^2 - Ax + \alpha + \frac{q}{2A}\right)\left(x^2 + Ax + \alpha - \frac{q}{2A}\right).$$

Solving the two quadratic equations yields the required formulas. $\square$

(b) **Theorem.** Let $p, q, s \in \mathbb{C}$, with $q \ne 0$. Let α denote any root of $q^2 = 4(2\alpha - p)(\alpha^2 - s)$. Let A denote any value of the square root of $2\alpha - p$. Then the roots of the equation $x^4 + px^2 + qx + s = 0$ are

$$\left(A + \sqrt{-2\alpha - p - \frac{2q}{A}}\right)/2 \quad \text{and} \quad \left(A + \sqrt{-2\alpha - p + \frac{2q}{A}}\right)/2,$$

where $\sqrt{y}$ is viewed as a multivalued function giving both root values of y; note that $A \ne 0$, because $q^2 = 4A^2(\alpha^2 - s) \ne 0$.

The proof is similar to the proof of the theorem from part (a).

3. Bezout's Theorem and its corollaries (2)

3.3.1. (a) Calculate the values of the functions

$$P(x) = 2x^3 - 27x^2 + 141x - 256 \ \text{ for } x = 16$$

and

$$Q(x) = x^4 + \frac{x^3}{4} - \frac{x^2}{2} + 1 \ \text{ for } x = -\frac{3}{4}.$$

Hint.

$$a_n x^n + a_{n-1} x^{n-1} + \ldots + a_1 x + a_0 = (\ldots((a_n x + a_{n-1})x + a_{n-2})x + \ldots + a_1)x + a_0.$$

This algorithm is called *Horner's method*.

(b) How many addition and multiplication operations do you need to calculate the value of a polynomial of nth degree? Compare the "conventional" way with Horner's method.

In order to understand the motivation for the definitions below, it is useful to have some experience with polynomial manipulation.

A *polynomial with real coefficients* is an infinite sequence $(a_0, \ldots, a_n, \ldots)$ of real numbers, among which there are only a finite number of nonzero numbers. The words "with real coefficients" are omitted in this section.

We associate a polynomial, that is, a sequence $P = (a_0, \ldots, a_n, \ldots)$, with the function $\overline{P} \colon R \to R$ given by the formula $\overline{P}(x) = a_0 + a_1 x + \ldots + a_n x^n + \ldots$ (the sum is finite). The polynomial $P = (a_0, \ldots, a_n, \ldots)$ is usually written in the form $P(x) = a_0 + a_1 x + \ldots + a_n x^n$, i.e., seemingly identical to $\overline{P}$. However, we will distinguish between P and $\overline{P}$, until we prove that they are "the same thing" (problem 3.3.5 (c)), or in those generalizations where they are "not the same thing" (problem 3.3.5 (f)).

3.3.2. (Challenge.)
Give definitions of
(a) the sum and product of polynomials;
(b) a polynomial with integer coefficients, a polynomial with rational coefficients, and a polynomial with coefficients in $\mathbb{Z}_p$.

The *degree of a polynomial* P (denoted by $\deg P$) is the largest number n such that $a_n \neq 0$. It is convenient to define the degree of the zero polynomial to be $-\infty$; in this case the following statements hold without the assumption that all polynomials are nonzero.

3.3.3. (Challenge.)
(a) The degree of the sum of polynomials of different degrees is equal to the largest of their degrees.
(b) The degree of the product of polynomials is equal to the sum of their degrees.

3.3.4. Let P be a nonzero polynomial and a be a real number.
(a) **Bezout's Theorem.** There exists a polynomial Q such that

$$P(x) = (x - a)Q(x) + P(a).$$

In other words, the polynomial $P(x) - P(a)$ *is divisible by* $(x - a)$. Moreover, $\deg Q < \deg P$.
(b) **Corollary.** If $P(a) = 0$, then there exists a polynomial Q such that $P(x) = (x - a)Q(x)$ and $\deg Q < \deg P$.
(c) For which values of a is the polynomial $x^{1000} + ax + 9$ divisible by $x + 1$?

A number x_0 is said to be a *root* of a polynomial P if $\overline{P}(x_0) = 0$.

3.3.5. (a) **Lemma.** If P is a polynomial and $a_1, \ldots, a_k$ are its different roots, then there exists a polynomial Q such that $P(x) = (x - a_1) \cdots (x - a_k)Q(x)$.

(b) **Lemma.** A polynomial of degree $n \geq 0$ has at most n roots.

(c) **Theorem.** If the values of two polynomials at all points are the same, then these polynomials are equal. In other words, if P and P_1 are polynomials and $P(x) = P_1(x)$ for all x, then $P = P_1$.

(d) **Theorem.** If the values of two polynomials of degree n coincide at $n + 1$ different points, then these polynomials are equal.

(e) Does the statement (c) hold if we assume the coefficients to be integers, or rational numbers, or elements of $\mathbb{Z}_p$?

(f) Does the statement (d) hold if we assume the coefficients to be integers, or rational numbers, or elements of $\mathbb{Z}_p$?

3.3.6. The following equalities hold for any pairwise different numbers a, b, c, d, and x:

(a)
$$\frac{c(x-a)(x-b)}{(c-a)(c-b)} + \frac{a(x-b)(x-c)}{(a-b)(a-c)} + \frac{b(x-c)(x-a)}{(b-c)(b-a)} = x;$$

(b)
$$\frac{d(x-a)(x-b)(x-c)}{(d-a)(d-b)(d-c)} + \frac{a(x-b)(x-c)(x-d)}{(a-b)(a-c)(a-d)}$$
$$+ \frac{b(x-d)(x-c)(x-a)}{(b-d)(b-c)(b-a)} + \frac{c(x-d)(x-b)(x-a)}{(c-d)(c-b)(c-a)} = x.$$

Suggestions, solutions, and answers

3.3.4. (a) *Hint.* First prove the statement for $P = x^n$. Then prove that if it is true for P and P', then it is true for $P + P'$ and bP for any number b.

3.3.5. (b) We prove the statement by induction on the degree n of P. The statement is true for $n = 0$: a polynomial of zero degree is a nonzero constant and hence has no roots. Suppose that *any nonzero polynomial Q of degree $k < n$ has at most k roots.* Consider an arbitrary nonzero polynomial P of degree n. Suppose that it has at least $n + 1$ distinct roots $x_0, x_1, x_2, \ldots, x_n$.

By the corollary in 3.3.4 (b), we have $P = (x - x_0)Q$ for some polynomial Q of degree less than n. Substituting $x = x_1$ into this equation yields $0 = (x_1 - x_0)Q(x_1)$, which implies that $Q(x_1) = 0$. Similarly, $x_2, x_3, \ldots, x_n$ are also roots of the polynomial Q, which contradicts the inductive hypothesis.

The same solution can be written in a more explicit form. Let the polynomial P of degree n have distinct roots $x_0, x_1, \ldots, x_n$. Rewrite it in the form

$$P(x) = b_n(x - x_1) \cdot \ldots \cdot (x - x_n) + b_{n-1}(x - x_1) \cdot \ldots \cdot (x - x_{n-1})$$
$$+ \cdots + b_1(x - x_1) + b_0$$

(this is *Newton's interpolation formula*). Successively substituting the numbers $x_1, x_2, \ldots, x_n, x_0$ into the equality $P(x) = 0$ yields $0 = b_0 = b_1 = \ldots = b_n$.

(e) The statement holds for polynomials with integer and rational coefficients, but does not hold for polynomials with coefficients in $\mathbb{Z}_p$. For example, consider the polynomials x^p and x.

4. Divisibility of polynomials (3*)
By A. Ya. Kanel-Belov and
A. B. Skopenkov

Let A and $B \neq 0$ be polynomials with real coefficients. We say that A is *divisible* by B if there exists a polynomial Q with real coefficients such that $A = BQ$. In this case we call B a *divisor* of A.

3.4.1. (a) Suppose that $A = BQ$ where A and $B \neq 0$ are polynomials with rational coefficients and Q is a polynomial with real coefficients. Then the coefficients of Q are also rational.

(b) Does a statement analogous to (a) hold if we replace rational coefficients with integers?

(c) Formulate definitions of divisibility for polynomials with integer, rational, and $\mathbb{Z}_p$ coefficients.

3.4.2. (a) Do there exist polynomials P and Q with integer coefficients such that P has no integer roots, P does not divide Q, and $P(n)$ divides $Q(n)$ for any integer n?

(b) If P and Q are polynomials with integer coefficients, P has no integer roots, the leading coefficient of polynomial P is equal to 1, and $P(n)$ divides $Q(n)$ for any integer n, then P divides Q.

(c)* Suppose that P and Q are nonzero polynomials with integer coefficients having no common divisors of positive degree. Then the sequence $\gcd(P(n), Q(n))$ contains finitely many values.

3.4.3. (a) **Theorem on division with remainder for polynomials.** For any two polynomials A and $B \neq 0$ with real coefficients, there exist unique polynomials Q and R with real coefficients such that $A = BQ + R$ with $\deg R < \deg B$. These polynomials are called the *quotient* and *remainder* of the division of A by B.

(b) Does the theorem hold for polynomials with integer, rational, or $\mathbb{Z}_p$ coefficients?

(c) Formulate and prove an analogous theorem for polynomials with integer coefficients whose leading coefficients are equal to 1.

(d) A polynomial has remainder 1 upon division by $x - 1$ and remainder -1 upon division by $x + 1$. What is the remainder when this polynomial is divided by $x^2 - 1$?

3.4.4. (a) Find at least one pair of polynomials U and V with rational coefficients such that

$$(2x^2 + x + 2)U(x) + (x^2 - 3x + 1)V(x) = 1.$$

(b) Remove the irrational quantity in the denominator of the fraction $\frac{1}{2\alpha^2 + \alpha + 2}$, where $\alpha^2 - 3\alpha + 1 = 0$.

(c) Remove the irrational quantity in the denominator of the fraction $\frac{1}{\alpha + 1}$, where $\alpha^3 - 3\alpha + 1 = 0$.

The Euclidean algorithm for polynomials is similar to the Euclidean algorithm for integers; see problem 1.5.9 (b).

3.4.5. A polynomial with coefficients in F is called *irreducible* over the set F if it cannot be factored into the product of two polynomials of lesser degrees with coefficients in F. Is factorization into irreducible polynomials unique for polynomials with (a) real; (b) integer; (c) rational; (d) $\mathbb{Z}_p$ coefficients?

3.4.6. If nonzero polynomials P and Q with integer coefficients have no common divisors of positive degree, then there exist $c_1 > c_2 > 0$ such that for any rational α, we have

$$c_2 h(\alpha)^n < h(P(\alpha)/Q(\alpha)) < c_1 h(\alpha)^n.$$

Here $n := \max(\deg P, \deg Q)$ and the *height* $h(p/q)$ of an irreducible fraction p/q is defined to be $\max(|p|, |q|)$ where $p \neq 0$, with $h(0) = 1$.

3.4.7.* Given a rectangle, cut off a square that shares the smaller of its sides. Perform the same procedure with the remaining rectangle, etc. Determine if the sequence of ratios of the sides of the rectangles is periodic if one of the sides of the original rectangle is 1 and the other is equal to

(a) $\sqrt{2}$;   (b) $(1 + \sqrt{5})/2$;   (c) $\sqrt[3]{2}$;   (d) $\sqrt{2005}$.

This problem involves the Euclidean algorithm for real numbers. For details see [**Arn16a**].

Hints and answers

3.4.3. (d) We have

$$P(x) = (x^2 - 1)Q(x) + ax + b \implies a + b = 1, \quad -a + b = -1.$$

3.4.4. (a) Use the Euclidean algorithm.

(b) Use part (a).

3.4.5. The solution is similar to problem 3.7.2.

Answers: (a, c) Yes; (b, d) No.

The uniqueness holds either if we view decompositions differing by a constant factor to be the same, or if we consider polynomials with integer coefficients and leading coefficient 1.

5. Applications of complex numbers (3*)

The author thanks O. E. Orel for useful discussions.

A complex number is a pair (a, b) of real numbers. It is written in the form $a+bi$. The *sum* of complex numbers is defined to be $(a+bi)+(a'+b'i) = (a + a') + (b + b')i$, and the *product* is defined to be $(a + bi)(a' + b'i) = (aa' - bb') + (ab' + a'b)i$. The formula for the product is engineered to ensure that the equality $i^2 = -1$ holds.

3.5.1. Represent each of the following in the form $a + bi$:

(a) $(1 + 2i)(2 - i) + (1 - 2i)(2 + i)$;   (b) $\frac{3+8i}{-5+2i}$;   (c) $\left(\frac{1-i}{1+i}\right)^3$;

(d) $\sqrt{3 - 4i}$.

3.5.2. Solve the following equations in complex numbers:

(a) $z^2 + 4z + 29 = 0$;   (b) $z^2 - (3 - 2i)z + 5 - 5i = 0$;   (c) $z^3 - 1 = 0$.

3.5.3. (a) For any complex number $z \neq 0$ there exists a complex number u such that $zu = 1$.

(b) A number $|a + bi| := \sqrt{a^2 + b^2}$ is called the *modulus* of the complex number $a + bi$. Prove that $|z_1 \cdot z_2| = |z_1| \cdot |z_2|$.

3.5.4. (a) **Trigonometric form of complex numbers.** For any complex number z, there exist real numbers $r \geq 0$ and φ such that $z = r(\cos \varphi + i \sin \varphi)$. Are the numbers r and φ unique?

(b) **Formula for product of complex numbers.** The following equality holds:

$$(\cos \varphi + i \sin \varphi)(\cos \psi + i \sin \psi) = \cos(\varphi + \psi) + i \sin(\varphi + \psi).$$

(c) **De Moivre's formula.** The following equality holds:

$$(\cos \varphi + i \sin \varphi)^n = \cos n\varphi + i \sin n\varphi.$$

(d) For any integer $n > 0$, solve the equation $z^n = 1$ in complex numbers.

3.5.5. Represent each of the following complex numbers in trigonometric form:

(a) $-1/2 + i\sqrt{3}/2$;   (b) $\sqrt{2} + \sqrt{2}i$;   (c) -5;   (d) $-17i$;

(e) $\sin \pi/6 + i \sin \pi/6$;   (f) $1 + \cos \varphi + i \sin \varphi$;   (g) $\frac{\cos \varphi + i \sin \varphi}{\cos \varphi - i \sin \varphi}$.

Other introductory material about complex numbers can be found, for example, in [**V+15**].

3.5.6. Factor the following polynomials into quadratic and linear polynomials with real coefficients:

(a) $x^4 + 4$;   (b) $x^4 + x^3 + x^2 + x + 1$;   (c) $x^n - 1$.

3.5.7. (a)* **Fundamental Theorem of Algebra.** *Any non-constant polynomial with complex coefficients has a complex root.*

(This statement can be used further without proof.)

(b) A polynomial with complex coefficients of degree n has exactly n roots, taking into account their multiplicity. It is said that a root z_0 of polynomial P has *multiplicity* k if P is divisible by $(z - z_0)^k$ and is not divisible by $(z - z_0)^{k+1}$.

(c) If $z_1, \ldots, z_n$ are roots of a polynomial P with leading coefficient a_n, each root occurring as many times as its multiplicity, then $P(z) = a_n(z - z_1) \cdots (z - z_n)$.

3.5.8. Define $\overline{a + bi} := a - bi$, called the *conjugate* of $a + bi$.

(a) The following equalities hold:

- $\overline{z + w} = \overline{z} + \overline{w}$;
- $\overline{zw} = \overline{z} \cdot \overline{w}$;
- $z \cdot \overline{z} = |z|^2$;
- $P(\overline{z}) = \overline{P(z)}$, for any polynomial P with real coefficients.

(b) Any polynomial with real coefficients can be factored into a product of polynomials of degrees 1 and 2 with real coefficients.

(c) If P is a polynomial with real coefficients and $P(x) > 0$ for any $x \in \mathbb{R}$, then there exist polynomials Q and R with real coefficients such that $P = Q^2 + R^2$.

3.5.9.* Find all polynomials with real coefficients such that $P(x^2 + x + 1) \equiv P(x)P(x + 1)$.

3.5.10. (a) (Challenge.) Express $\cos n\varphi$ and $\sin n\varphi$ in terms of $\cos \varphi$ and $\sin \varphi$.

(b) **Lemma.** One can express $\cos n\varphi$ and $\frac{\sin n\varphi}{\sin \varphi}$ as polynomials in $\cos \varphi$.

3.5.11. Find $\{x_n\}$ and $\{y_n\}$ if $\begin{cases} x_{n+1} = 3x_n - 4y_n, \\ y_{n+1} = 3y_n + 4x_n, \end{cases}$
and (a) $x_0 = 1$, $y_0 = 0$;   (b) $x_0 = 1$, $y_0 = 2$.

3.5.12. Find (a) $\displaystyle\sum_{k=0}^{n} \cos k\varphi$;   (b) $\displaystyle\sum_{k=0}^{n} 2^k \sin k\varphi$;   (c) $\displaystyle\sum_{k=0}^{\infty} \frac{\cos k\varphi}{3^k}$.

3.5.13. (a) For $0 < x < \pi/2$, the following inequality holds:

$$\cot^2 x < \frac{1}{x^2} < \cot^2 x + 1.$$

(b) For any $k = 1, \ldots, n$ the following equality holds:

$$\sum_{j=0}^{n} (-1)^j \binom{2n+1}{2j+1} \cot^{2n-2j} \frac{\pi k}{2n+1} = 0.$$

(c) $\displaystyle\sum_{k=1}^{n} \cot^2 \frac{\pi k}{2n+1} = \frac{n(2n-1)}{3}$.

(d) $\displaystyle\sum_{k=1}^{\infty} \frac{1}{k^2} = \frac{\pi^2}{6}$.

(e, f)* Find $\displaystyle\sum_{k=1}^{\infty} \frac{1}{k^4}$ and $\displaystyle\sum_{k=1}^{\infty} \frac{1}{k^6}$.
The infinite sums used here are defined in section 5.

Hints and answers

3.5.6. Find all complex roots of these polynomials.

3.5.8. (b) By (a), the complex roots of this polynomial can be grouped into conjugate pairs.
     (c) The product of two sums of squares is also a sum of squares.

3.5.11. Take $z_n = x_n + iy_n$.

3.5.12. *Hint.* Write $\mathrm{Re}(a + bi) := a$ for real a and b. Use the fact that $\cos k\varphi = \mathrm{Re}(\cos \varphi + i \sin \varphi)^k$.

3.5.12. *Answers*: (a) $\dfrac{\sin \frac{n+1}{2}\varphi \cos \frac{n}{2}\varphi}{\sin \frac{\varphi}{2}}$. (b) $\dfrac{2^{n+2} \sin n\varphi - 2^{n+1} \sin(n+1)\varphi + 2 \sin \varphi}{5 - 4\cos \varphi}$.

     (c) $\dfrac{9 - 3\cos \varphi}{10 - 6\cos \varphi}$.

3.5.13. (a) Use the fact that $\sin x < x < \tan x$ for $0 < x < \pi/2$.
     (b) Note that $\left(\cos \frac{\pi k}{2n+1} + i \sin \frac{\pi k}{2n+1} \right)^{2n+1} = (-1)^k$.

6. Vieta's Theorem and symmetric polynomials (3*)

3.6.1. (a) Construct the polynomial whose roots are cubes of the roots of
the equation $x^2 - 6x + 6 = 0$.

(b) Express $x^3 + 4x^2y + 4xy^2 + y^3$ in terms of $x + y$ and xy.

(c) Solve the following system of equations:

$$\begin{cases} x^3y + xy^3 = 300, \\ xy + x^2 + y^2 = 37. \end{cases}$$

3.6.2. (a, b, c) Represent

$$x^2 + y^2 + z^2, \quad x^2y + y^2z + z^2x + x^2z + z^2y + y^2x, \quad x^3 + y^3 + z^3$$

as polynomials in

$$\sigma_1 := x + y + z, \quad \sigma_2 := xy + yz + zx, \quad \text{and} \quad \sigma_3 := xyz.$$

(d) Is it possible to represent $(x^{100}y + y^{100}z + z^{100}x)(x^{100}z + z^{100}y + y^{100}x)$
as a polynomial in σ_1, σ_2, and σ_3?

Formulate your own definition of a polynomial in several variables and
its multi-degree. Generalizing the notation above, we define the *elementary
symmetric polynomials* $\sigma_1, \sigma_2, \ldots, \sigma_n$ by

$$\sigma_k := \sum_{1 \le i_1 < i_2 < \cdots < i_k \le n} x_{i_1} x_{i_2} \cdots x_{i_k},$$

where the number of variables is n. For example, if $n = 4$, then

$$\sigma_2 = x_1x_2 + x_1x_3 + x_1x_4 + x_2x_3 + x_2x_4 + x_3x_4.$$

3.6.3. (a) The multi-degree of the product of polynomials in several variables is equal to the sum of their multi-degrees.

(b) A polynomial f in two variables x and y is called *symmetric* if
the polynomials $f(x, y)$ and $f(y, x)$ are equal. Prove that any symmetric
polynomial in two variables x and y can be expressed as polynomial in $x + y$
and xy.

(c) A polynomial f in n variables $x_1, x_2, \ldots, x_n$ is called symmetric if
$f(x_1, x_2, \ldots, x_n) = f(x_{\sigma(1)}, x_{\sigma(2)}, \ldots, x_{\sigma(n)})$ for every permutation σ of the
set $\{1, 2, \ldots, n\}$.

Prove that any symmetric polynomial in three variables x, y, and z can
be expressed as a polynomial in σ_1, σ_2, and σ_3.

(d) **Fundamental Theorem on Symmetric Polynomials.** Any symmetric polynomial in n variables can be expressed as a polynomial in the
elementary symmetric functions σ_1, $\sigma_2, \ldots$, and σ_n.

3.6.4. Let $x_1 < x_2 < \ldots < x_7$, $y_1 < y_2 < \ldots < y_7$, $x_1 < y_1$, and $\sum_{i=1}^{7} x_i^k = \sum_{i=1}^{7} y_i^k$ for any $k \in \{1, \ldots, 6\}$. Then $x_7 < y_7$.

The following important result is a simple consequence of factoring a polynomial into terms of the form $(x-r)$, where r is a root of the polynomial (cf. Lemma 3.3.5 (a)).

Theorem 3.6.5 (Vieta's Theorem). Let $x_1, x_2, \ldots, x_n$ be the roots of the polynomial
$$x^n + a_1 x^{n-1} + a_2 x^{n-2} + \cdots + a_n = 0.$$
Then $a_k = (-1)^k \sigma_k$ for $k = 1, 2, \ldots, n$.

Suggestions, solutions, and answers

3.6.1. (c) See [**Vin80**, IX.2.6, ex. 1].

3.6.2. *Answers*: (a) $\sigma_1^2 - 2\sigma_2$; (b) $\sigma_1 \sigma_2 - 3\sigma_3$; (c) $\sigma_1^3 - 3\sigma_1\sigma_2 + 3\sigma_3$.
(d) Use 3.6.3 (c).

3.6.3. (b) Use induction on the multi-degree of the polynomial, in lexicographic order. For a symmetric polynomial f of multi-degree (k, l), i.e., with leading term $ax^k y^l$, $k \geq l$, consider the polynomial $f - a(x + y)^{k-l}(xy)^l$.

(c) Use induction; see (b). For the symmetric polynomial f of multi-degree (k, l, m), consider the polynomial $f - a\sigma_1^{k-l}\sigma_2^{l-m}\sigma_3^m$.

(d) To prove the Fundamental Theorem on Symmetric Polynomials we use induction on the multi-degree of the given symmetric polynomial $f(x_1, x_2, \ldots, x_n)$ in lexicographic order. The base case $f = 0$ is obvious.

To prove the inductive step, denote the lexicographically leading term of the polynomial f by $u := ax_1^{k_1} x_2^{k_2} \cdots x_n^{k_n}$.

Suppose $k_i < k_{i+1}$ for some i. Then together with u, the polynomial must contain a term $ax_1^{k_1} \cdots x_i^{k_{i+1}} x_{i+1}^{k_i} \cdots x_n^{k_n}$, which comes before u in lexicographic order, a contradiction. Therefore, $k_1 \geq k_2 \geq \ldots \geq k_n$. According to (a), the leading term of the polynomial $g := a\sigma_1^{k_1-k_2}\sigma_2^{k_2-k_3} \cdots \sigma_{n-1}^{k_{n-1}-k_n}\sigma_n^{k_n}$ coincides with u. So the multi-degree of polynomial $f - g$ is less than the multi-degree of polynomial f. We now apply the inductive hypothesis to $f - g$. $\qquad\square$

7. Diophantine equations and Gaussian integers (4*)
By A. Ya. Kanel-Belov

Everybody knows the Euclidean algorithm well. Given two numbers a and b, the greater of them is selected, the smaller is subtracted from the larger, the larger is replaced by the difference, and with the new pair of numbers the same procedure is performed again. See problem 1.5.9 (b). We used the Euclidean algorithm earlier to prove various properties of integers, e.g., in sections 5 and 4 We shall now demonstrate a novel (to most readers) application of the Euclidean algorithm.

3.7.1. Solve the following equations in integers:

$$\text{(a) } x^2 + 4 = y^3; \quad \text{(b) } x^2 + 2 = y^n; \quad \text{(c)}^* \; x^3 + y^3 = z^3.$$

Try to solve them without reading further! However, you are unlikely to succeed. Return to them after you have read this section.

When confronted with $x^2 + 4 = y^3$ in integers, you probably considered the factorization $x^2 + 4 = (x + 2i)(x - 2i)$. For odd x, the two factors are relatively prime and therefore both must be cubes. This leads to a solution. (When x is even, it's trickier: both factors are divisible by $(1 + i)^3$. Try to solve the equation and then compare your solution with the one at the end of the section.)

The idea is that we benefit from the additional possibilities in the factorization due to the use of *Gaussian integers*, i.e., numbers of the form $a + bi$ with integer a and b. However, since life is not a bowl of cherries, it only works sometimes (see problems 1.2.8 (b) and 3.7.3 (b)). In order to use factorization to solve equations, we need *uniqueness of factorization into primes*. This would allow us to inherit all the nice arithmetical properties of the integers. The following problem illuminates an amazing phenomenon: to get the *arithmetical* goodies, it is sufficient to prove a *geometric* property about the possibility of division with a remainder.

3.7.2. A Gaussian integer is called *prime* if it cannot be decomposed into two Gaussian factors, each different from ± 1 and $\pm i$.

(a) The uniqueness of factorization into prime factors is a consequence of the following analogue of Euclid's lemma 1.5.7 (c):

Generalized Euclid's lemma. For any a and b, if a prime p divides ab, then p divides a or p divides b.

(b) The generalized Euclid's lemma is a consequence of the following fact (an analogue of the lemma about representation of the GCD in 1.5.7 (a)):

Principal ideal property. For any a and b there exist x and y such that $xa + yb = \gcd(a, b)$. Give your own definition of the greatest common divisor $\gcd(a, b)$ of Gaussian integers a and b.

(c) The principal ideal property results from the following property (an analogue of the theorem about division with a remainder in 1.4.1 (b)):

Euclidean property. For any $b \neq 0$ and a there exists k such that $|a - kb| < |b|$.

3.7.3. Is the Euclidean property (and, therefore, the unique factorization property) true for the set $Z[\xi]$ of numbers of the form $a + b\xi$ where a and b are integers, if ξ is

(a) $\sqrt{-2}$;   (b) $\sqrt{-3}$;   (c) $(1 - \sqrt{-3})/2$;   (d) $(1 - \sqrt{-5})/2$;
(e) $(1 - \sqrt{-7})/2$?

3.7.4. (a) No prime of the form $4k - 1$ can be expressed as a sum of two squares.

(b) Any prime of the form $4k + 1$ can be expressed as a sum of two squares exactly in one way.

(c) There exists an integer which can be decomposed into the sum of squares in exactly 1024 ways.

This problem is easier to solve without Gaussian integers (see section 3), but it is instructive to practice using them!

See more in [**Pos78**, §4]. See also problem 3.4.7.

Suggestions, solutions, and answers

3.7.1. (a) (By R. I. Devyatov) *Answer*: $x = \pm 2$, $y = 2$ and $x = \pm 11$, $y = 5$.

Pass to Gaussian integers and obtain $(x + 2i)(x - 2i) = y^3$.

A Gaussian integer is called a *perfect cube* if it is equal to b^3 for some Gaussian integer b. Note that all invertible numbers ± 1 and $\pm i$ are perfect cubes. So all Gaussian integers of the form ωa^3, where a is a Gaussian integer and ω is one of the invertible numbers ± 1 and $\pm i$, are perfect cubes.

Two Gaussian integers a and b are called *associates* if $a = \omega b$ where ω is one of the invertible numbers ± 1 and $\pm i$.

Lemma. Both $x + 2i$ and $x - 2i$ are perfect cubes.

Proof. We use the uniqueness of decomposition into prime Gaussian factors up to multiplication by invertible numbers ± 1 or $\pm i$. Let $d := \gcd(x+2i, x-2i)$. Then $x + 2i - (x - 2i) = 4i = -i(1 + i)^4$ is divisible by d. Since $1 + i$ is prime, d is a power of $(1 + i)$ of degree at most four.

Note that the decomposition of $x - 2i$ into prime Gaussian integers can be obtained from the decomposition of $x + 2i$ by replacing all factors by their conjugates.

Since $1 + i = i(1 - i)$, the powers to which $(1 + i)$ occurs in the decomposition of $x + 2i$ and $x - 2i$ into prime factors are the same. Denote this power by k.

Then y^3 is divisible by $(1 + i)^{2k}$. Since $2k$ is divisible by 3, k is also divisible by 3. Since d is a power of $1 + i$ of degree at most four, the Gaussian integer $x + 2i$ either is not divisible by $1 + i$ or is divisible by $(1 + i)^3$.

If $x + 2i$ is not divisible by $1 + i$, then $x + 2i$ and $x - 2i$ are relatively prime. Since their product is a perfect cube, $x + 2i$ and $x - 2i$ must both be perfect cubes.

If $x + 2i = a(1+i)^3$ for some Gaussian integer a, then $x - 2i = b(1+i)^3$ for some Gaussian integer b. Thus $y^3 = ab(1+i)^6$. So $ab = \left(\frac{y}{(1+i)^2}\right)^3$ is a perfect cube. Since a and b are relatively prime, they are perfect cubes. Therefore $x + 2i = a(1+i)^3$ and $x - 2i = b(1+i)^3$ are perfect cubes, proving the lemma. $\square$

Continuation of solution. Write

$$x + 2i = (c + di)^3 = c^3 + 3c^2 di + 3cd^2 i^2 + d^3 i^3 = c^3 - 3cd^2 + (3c^2 d - d^3)i.$$

Compare the imaginary parts: $2 = 3c^2 d - d^3 = d(3c^2 - d^2)$. This is an equality of ordinary integers, so $d = \pm 2$ or $d = \pm 1$.

Case 1: $d = \pm 1$. Then $3c^2 - 1 = \pm 2$, i.e., $3c^2 = -1$ or 3. It cannot be equal to -1, so $c = \pm 1$, $c + di = 1 + i$ or an associate of it, and $x + 2i = 2 + 2i$ or one of its associates. Therefore $x = \pm 2$ and $y = 2$.

Case 2: $d = \pm 2$. Then $3c^2 - 2 = \pm 1$, i.e., $3c^2 = 1$ or 3. It cannot be equal to 1, so $c = \pm 1$, $c + di = 2 + i$ or one of its associates, and $x + 2i = 11 + 2i$ or one of its associates. Thus $x = \pm 11$ and $y = 5$.

3.7.1. (b) Use problem 3.7.3 (a).

(c) Use problem 3.7.3 (c).

3.7.2. (a) If p does not divide b, then $pm + bn = 1$. If at the same time p divides ab, then p divides $nab + mpa$; i.e., p divides a.

(b) Divide a by b with remainder a': $a = kb + a'$. Any common divisor of a' and b is a common divisor of a and b. Likewise, the set of Gaussian integers of the form $ma' + nb$ contains a and certainly contains b, and therefore contains any Gaussian integer of the form $pa + qb$. Similarly, we can verify the converse statement: the set of Gaussian integers of the form $pa + qb$ contains the set of Gaussian integers of the form $ma' + nb$. Thus the pair (a, b) may be replaced by the pair (a', b), which in a sense is "smaller." The process stops at the pair $(\gcd(a, b), 0)$. The details of this proof are similar to the proof of the GCD representation lemma; see problem 1.5.7 (a).

(c) The set of Gaussian integers $(p + qi)b$, that is, multiples of b, forms a lattice of squares with side $|b|$. The Gaussian integer a falls into one of the lattice squares. It suffices to apply the geometric fact that *the distance from any point inside a square to the nearest vertex is strictly less than the length of the side of the square.*

3.7.3. (a) *Answer:* The unique factorization property holds.

The proof is similar to that of problem 3.7.2 (c). A necessary geometric fact is that *the distance from any point inside a $\sqrt{2} \times 1$ rectangle to the nearest vertex is strictly less than 1.*

(b) *Answer*: The unique factorization property does not hold.
Example: $4 = 2 \cdot 2 = (1 + \sqrt{-3})(1 - \sqrt{-3})$. Think about why the corresponding geometric fact is incorrect.

8. Diagonals of regular polygons (4*)
By I. N. Shnurnikov

The goal is to determine which diagonals of a regular n-gon and how many of them can intersect at one point. Problem 3.8.2 describes possible intersection points, and problems 3.8.4 and 3.8.6 are needed to prove the impossibility of other points of intersection, which ends with a computer-assisted analysis of cases.

3.8.1. (a) In an isosceles triangle ABC with base BC, the vertex angle A is $80°$. Inside the triangle a point M is chosen so that $\angle MBC = 30°$ and $\angle MCB = 10°$. Then $\angle AMC = 70°$.

(b) Choose P inside square $ABCD$ so that triangle ABP is equilateral. Then $\angle PCD = 15°$.

(c) In an isosceles triangle ABC with base AC, the angle at the vertex B is equal to $20°$. On sides BC and AB, choose points D and E respectively so that $\angle DAC = 60°$ and $\angle ECA = 50°$. Then $\angle ADE = 30°$.

In this section, the word "intersect" means "intersect at a single point."

3.8.2. (a) Diagonals A_1A_{n+2}, $A_{2n-1}A_3$, and $A_{2n}A_5$ of a regular $2n$-gon intersect.

(b) Diagonals A_1A_7, A_3A_{11}, A_4A_{16} and A_5A_{21}, of a regular 24-gon intersect.

(c) In the regular 30-gon, the following seven diagonals intersect:

$$A_1A_{13}, \ A_2A_{17}, \ A_3A_{21}, \ A_4A_{24}, \ A_5A_{26}, \ A_8A_{29}, \ A_{10}A_{30}.$$

3.8.3. (a) Let ABC be a triangle with $\angle A = 50°$, $\angle B = 60°$, and $\angle C = 70°$. Choose points D and E respectively on sides BA and BC so that $\angle DCA = 50°$ and $\angle EAC = 40°$. Then $\angle AED = 30°$.

(b) Let ABC be a triangle with $\angle A = 14°$, $\angle B = 62°$ and $\angle C = 104°$. On sides AC and AB choose points D and E respectively so that $\angle DBC = 50°$ and $\angle ECB = 94°$. Then $\angle CED = 34°$.

3.8.4. If p is prime, then no three diagonals of the regular p-gon intersect at one point in the interior.

Theorem 3.8.5 ([PR98]). For $n > 4$, the maximum number of diagonals of a regular n-gon that intersect at one point (other than the center or a vertex) is equal to:

2 if n is odd or $n = 6$;

3 if n is even and not divisible by 6;

4 if $n = 12$;

5 if n is divisible by 6, n is not divisible by 30, and $n \notin \{6, 12\}$;

7 if n is divisible by 30.

3.8.6. a) Let p be prime and let S be a polynomial of degree not greater than $2p - 1$ with integer coefficients that has the root $e^{\frac{i\pi}{p}}$. Then

$$S(x) = a(1 + x^2 + x^4 + \ldots + x^{2p-2}) + \sum_{j=0}^{p-1} a_j(x^j + x^{p+j})$$

for some $a, a_0, a_1, \ldots, a_{p-1} \in \mathbb{Z}$.

(b) A nonzero polynomial $S(x) = \sum_{j=1}^{k} a_j x^j$ with non-negative integer coefficients is called k-minimal if $S(e^{\frac{2\pi i}{k}}) = 0$ and there do not exist integers $b_j,\ 0 \le b_j \le a_j$, such that $\sum_{j=1}^{k} b_j e^{\frac{2\pi i j}{k}} = 0$, where not all of the b_j are equal to zero and not all of the b_j are equal to a_j. Prove that for every k-minimal polynomial S, there exist distinct primes $p_1 < p_2 < \ldots < p_s \le k$, integers m and l, and a $p_1 p_2 \cdots p_s$-minimal polynomial S_1 satisfying $S(x) = x^l \cdot S_1(x^m)$.

(c) For a k-minimal polynomial S, choose $m, l, p_1, p_2, \ldots, p_s$ and S_1 from (b) above with minimal p_s. Suppose $p_1 = 2$ and $S(1) < 2p_s$. Then there exist integers $l, r < p_s$ and $p_1 p_2 \cdots p_{s-1}$-minimal polynomials $T_1, T_2, \ldots T_r$ satisfying

$$S(x) = x^l \cdot \sum_{j=1}^{r} T_j^j(x) \quad \text{and} \quad \sum_{j=1}^{r} T_j(1) = 2r + S(1) - p_s.$$

(d) There exist exactly 107 k-minimal polynomials with $k > 0$ whose values at 1 do not exceed 12.

Suggestions, solutions, and answers

3.8.1. Perform additional constructions and reduce the problem to finding intersections of diagonals in a regular n-gon.

3.8.2. Use the trigonometric form of Ceva's Theorem. (See [**Cev**] for a nice discussion of Ceva's Theorem and its trigonometric form.)

3.8.3. Using isogonal conjugates [**ISO**], the problem is reduced to previous ones.

3.8.4, 3.8.5. If you cannot solve them, continue reading.

Reformulate the trigonometric form of Ceva's theorem for the point of intersection of three distinct diagonals of an n-gon into the equation $\sum_{j=1}^{6} e^{i\pi x_j} + \sum_{j=1}^{6} e^{-i\pi x_j} = 0$, where the six values $x_j, j = 1, 2, \ldots, 6$, are defined by a certain formula (which should be found) and satisfy the equality $\sum_{j=1}^{6} x_j = 1$.

9. A short refutation of Borsuk's conjecture

This section[2] provides the simplest known refutation of Borsuk's conjecture: *Any bounded subset of n-dimensional Euclidean space containing more than n points can be partitioned into n + 1 non-empty sets of smaller diameter.* The presented counterexample is due to N. Alon and is a wonderful application of combinatorics and algebra to geometry.

Theorem 3.9.1 (Borsuk). Any bounded subset of the plane that contains more than two points can be partitioned into three non-empty sets of smaller diameter.

The *diameter* of a non-empty subset of a plane is the greatest distance between its points (more precisely, the supremum of these distances). A subset of a plane is called *bounded* if its diameter is finite. (For subsets of n-dimensional Euclidean space, these terms have analogous definitions.)

The diameter of the empty set is assumed to be zero.

Borsuk conjectured a higher-dimensional generalization of his result, which for many years was one of the most intriguing problems of combinatorial geometry.

A *point* $x = (x_1, \ldots, x_n)$ in n-dimensional Euclidean space is an ordered set of n real numbers. The *distance* between points $x = (x_1, \ldots, x_n)$ and $y = (y_1, \ldots, y_n)$ is given by the formula

$$|xy| = \sqrt{(x_1 - y_1)^2 + \cdots + (x_n - y_n)^2}.$$

The *diameter* and the property of being *bounded* for a subset of the n-dimensional Euclidean space are defined in the same way as for a subset of the plane.

This conjecture states that any bounded subset of n-dimensional Euclidean space containing more than n points can be divided into $n + 1$ non-empty parts of smaller diameter.

[2] The author thanks N. Dolbilin and A. Raygorodsky, from whom he learned about counterexamples to the hypothesis of Borsuk, students of Moscow School 57, who learned these counterexamples from him, and M. Akhmedov, V. Dubrovsky, I. Pak, and A. Rukhovich for helpful discussions.

It is not difficult to construct a subset of n-dimensional Euclidean space which cannot be divided into n parts of smaller diameter. For $n = 3$ it is the regular tetrahedron; for any n it is the n-dimensional simplex.

In 1993 J. Kahn and G. Kalai used combinatorial ideas of Boltyanski, Erdős, and Larman to find a counterexample to Borsuk's conjecture; see [**KBK08**]. A detailed history of the problem is described in [**AZ04**, **Rai04**].

Theorem 3.9.2. There exist an integer n and a bounded subset M of n-dimensional Euclidean space containing more than n points such that M cannot be divided into $n + 1$ parts of smaller diameter.

We will present the simplest known proof, due to N. Alon; cf. [**Nil94**, **Skob**, **Ger99**, **AZ04**, **Rai04**]; other proofs give stronger results. It is an amazing example of an important result in modern mathematics that does not require two years of prerequisite courses followed by a semester-long special course for full comprehension. Other simple examples using similar algebraic techniques in combinatorics can be found in [**RSG+16**, 7.1].

Proof. Let

$$M = \{(x_1, \ldots, x_n) \text{ such that } x_1 = 1, \, x_2, \ldots, x_n \in \{1, -1\},$$
$$\text{and an odd number of coordinates } x_2, \ldots, x_n \text{ are equal to } 1.\}$$

Each vertex of an n^2-dimensional cube is an ordered n^2-tuple of 1's and -1's. It is convenient to think of it as a $n \times n$ matrix. However, if you prefer, imagine a vector with n^2 elements.

We will map each point $x = (x_1, \ldots, x_n) \in M$ to a matrix $x^T \otimes x$, defined by $(x^T \otimes x)_{ij} := x_i x_j$. For example,

$$(1, -1, -1)^T \otimes (1, -1, -1) = \begin{pmatrix} 1 & -1 & -1 \\ -1 & 1 & 1 \\ -1 & 1 & 1 \end{pmatrix}$$

$$= (1, -1, -1, -1, 1, 1, -1, 1, 1).$$

We will prove that the set

$$M' = \{x^T \otimes x : x \in M\}$$

provides a counterexample to Borsuk's conjecture, for $n = 4p$ where p is a sufficiently large prime.

Let $x, y \in M$. Then

$$(x_i x_j - y_i y_j)^2 = (x_i x_j)^2 (1 - x_i y_i x_j y_j)^2 = (1 - x_i y_i x_j y_j)^2.$$

Let $a = a(x, y)$ denote the number of indices i for which $x_i = y_i$. Then $x_i y_i = 1$ for a indices i and $x_i y_i = -1$ for $n - a$ indices i. Thus $|x^T \otimes x, y^T \otimes y|^2 = 4a(n - a)$. This expression reaches its maximum at $a = n/2$. Consequently, the condition $|x^T \otimes x, y^T \otimes y| = \operatorname{diam} M'$ is equivalent to $a = n/2$.

Therefore, if the set M' is partitioned into k sets $Z'_1, \ldots, Z'_k$ of smaller diameter, then in every subset Z_j of M corresponding to one part Z'_j, no two vectors differ at exactly half of the coordinates. Since for any $x \in M$, $x_1 = 1$, we have $x^T \otimes x \neq y^T \otimes y$ for $x \neq y$. Thus $|Z_i| = |Z'_i|$. The theorem follows from the estimation lemma 3.9.3 below since $|M| = 2^{n-2}$. □

Lemma 3.9.3 (Estimation). If p is a sufficiently large prime and $n = 4p$, then among any $\lfloor 2^{n-2}/(n^2 + 1) \rfloor$ vectors in M there are two that differ at exactly half of the coordinates.

When proving Lemma 3.9.3, we do not need to remember the construction $x^T \otimes x$.

3.9.4. For a prime p and an integer t, the number
$$G(t) := (t - 1)(t - 2) \cdots (t - p + 1)$$
is divisible by p if and only if t is not divisible by p.

Any polynomial $\lambda_1 F_1 + \ldots + \lambda_s F_s$ with rational $\lambda_1, \ldots, \lambda_s$ is called a *rational linear combination* of polynomials $F_1, \ldots, F_s$. For example, the polynomial x_2 is a rational linear combination of the polynomials $2x_1$, 1, and $x_1 + x_2$.

Polynomials are called *linearly independent* if the only rational linear combination of them that equals zero requires all λ_k to equal zero. For example, the polynomials $1, x_2, x_3, \ldots, x_n$ are linearly independent.

(*) A polynomial in $n - 1$ variables $x_2, \ldots, x_n$ with rational coefficients *has degree less than $n/4$* if it is a rational linear combination of polynomials of the form $x_2^{\alpha_2} \cdots x_n^{\alpha_n}$, where $\alpha_2, \ldots, \alpha_n$ are non-negative integers whose sum is less than $n/4$.

The estimation lemma 3.9.3 follows from the linear independence lemma 3.9.5 below and from statement 3.9.6.

Lemma 3.9.5 (Linear independence). Let p be prime, $n = 4p$, and $A \subset M$ such that no two vectors in A differ at exactly half of the coordinates. For each vector $a \in A$, define the polynomial F_a in $x_2, \ldots, x_n$ with coefficients in $\mathbb{Z}_p$ by
$$F_a(x_2, \ldots, x_n) := G(a \cdot (1, x_2, \ldots, x_n)).$$
Then the polynomials F_a, $a \in A$, all have degrees less than $n/4$ and are linearly independent.

3.9.6. Let q be prime and n be a sufficiently large integer (note that $n/4$ need not be prime nor be equal to q). Then any family of polynomials in $x_2, \ldots, x_n$ with coefficients in $\mathbb{Z}_q$ of degree less than $n/4$ that is linearly independent over $\mathbb{Z}_q$ contains fewer than $\lfloor 2^{n-2}/(n^2 + 1) \rfloor$ polynomials.

3.9.7.[*] Borsuk's conjecture is false for
   (a) $d = 946$ [**Nil94**];   (b) $d = 561$ [**Rai04**].

Suggestions, solutions, and answers

3.9.1. First, use continuity to prove that any planar figure of diameter 1 can be positioned inside a regular hexagon whose inscribed circle has diameter 1. Then prove that although the obtained regular hexagon has diameter greater than 1, it can be cut into three pieces of diameter less than 1 (cf. [**Yan**]).

3.9.5.

Proof of the linear independence lemma. The statement about the degree is obvious. To prove linear independence, assume to the contrary that $\lambda_1 F_{a_1} + \cdots + \lambda_s F_{a_s} = 0$ for some $a_1, \ldots, a_s \in A$ with $\lambda_1, \ldots, \lambda_s \in \mathbb{Z}_p$ and not all λ_k zero. Here the $a_1, \ldots, a_s$ are vectors, not scalar coordinates. Without loss of generality, we can assume that $\lambda_1 \neq 0 \in \mathbb{Z}_p$. In the above equality, for each $j = 2, \ldots, n$ take $x_2 = (a_1)_2, \ldots, x_n = (a_1)_n$.

Recall that the dot product of vectors is an integer and not a residue modulo p. From the equality $a_1 \cdot a_1 = n = 4p$ and assertion 3.9.4 it follows that $\lambda_1 F_{a_1} \neq 0$. If $a = b$ then $a \cdot b = a \cdot a = n$ is divisible by 4. For each $a, b \in A$, replacing in a (or in b) two 1's by -1's does not change $a \cdot b \bmod 4$. For each $a \in A$ the number of 1's in a is even. Then $a \cdot b$ is divisible by 4 for each $a, b \in A$. Therefore $a \cdot b \notin \{\pm p, \pm 2p, \pm 3p\}$. Also, $a \cdot b \neq n = 4p$ because $a \neq b$, and $a \cdot b \neq -n$ because the first coordinates of a and of b are both equal to 1.

From this and the fact that $a \cdot b \neq 0$ it follows that $a \cdot b$ is not divisible by p. Thus 3.9.4 implies that $\lambda_k F_{a_k} = 0$ for any $k > 1$, a contradiction. $\square$

3.9.6.

Proof. The number of ordered solutions $(\alpha_2, \ldots, \alpha_n)$ to the equation $\alpha_2 + \ldots + \alpha_n = d$ in non-negative integers is equal to $\binom{n+d-2}{d}$.

For $d < p := \lfloor n/4 \rfloor$, we have

$$\binom{n+d-2}{d} \overset{(1)}{<} \binom{n+p-3}{p-1} \overset{(2)}{<} \binom{5p}{p-1} \overset{(3)}{<} \frac{(4+1)^{5p}}{4^{4p+1}} \overset{(4)}{<} 13^p.$$

Here
   - inequality (1) holds because $2d < 2p < n - d - 2$;
   - inequality (2) holds because $n + p - 3 < 5p$;
   - inequality (3) follows from the Newton binomial formula for $(4+1)^{5p}$ (cf. [**RSG+16**], problem 6.1.5);
   - inequality (4) holds because $5^5 < 2^7 \cdot 5^2 < 2^8 \cdot 13$.

Since $13 < 2^4$, this implies that for sufficiently large n the number r of polynomials in the family (*) does not exceed $n \cdot 13^{n/4} < \lfloor 2^{n-2}/(n^2 + 1) \rfloor$.

Let $Q_1, \ldots, Q_r$ denote the family of polynomials (*) and let $F_1, \ldots, F_k$ be the given linearly independent family. Consider the $k \times r$ matrix of elements $\lambda_{i,j} \in \mathbb{Z}_q$ for which $F_i = \sum_j \lambda_{i,j} Q_j$ for any $i = 1, \ldots, k$. The family of polynomials obtained from the family $F_1, \ldots, F_k$ by replacing F_i with $F_i + \lambda F_j$, $j \neq i$, is also linearly independent. By such substitutions and permutations of polynomials (i.e., by Gaussian elimination of the unknowns), the $k \times r$ matrix can be reduced to "upper triangular" form. Since the polynomials $F_1, \ldots, F_k$ are linearly independent, there is no zero row in this new matrix. Thus $k \leq r$. $\qquad \square$

3.9.7. (a) We built our example in an $\frac{n(n-1)}{2}$-dimensional space (albeit with a different metric) whose points are given by sets z_{ij} in which the indices i and j run from 1 to n so that $i < j$.

Prove that for each $k \leq 7$ we have $\binom{27}{k}\left(\frac{28 \cdot 27}{2} + 1\right) < 2^{26}$.

(b) Similar to (a), only $x_1 = x_2 = x_3 = 1$, $n = 36$, and $G(t) = \frac{1}{9}(t - 1)(t - 2)(t - 3)(t - 5)(t - 6)(t - 7)(t - 8)$.

Chapter 4

Permutations

Solving the problems in this chapter does not require any prior knowledge. The problems relate more to combinatorics than to algebra up until their connection with the solution of equations is explored (see Chapter 8). They naturally lead the reader to the concept of a group of transformations, which is explicitly introduced in subsection3.I of Chapter 8. A mini-course on group theory can be constructed from this chapter, Chapter 2 "Multiplication modulo a prime number", and Chapter 8 "Solvability in radicals".

1. Order, type, and conjugacy (1)

4.1.1. Fifteen students sit on fifteen numbered chairs. Every minute a kind teacher moves them according to the following scheme:

$$\begin{pmatrix} 1 & 2 & 3 & 4 & 5 & 6 & 7 & 8 & 9 & 10 & 11 & 12 & 13 & 14 & 15 \\ 3 & 5 & 10 & 8 & 11 & 14 & 15 & 6 & 13 & 1 & 4 & 9 & 7 & 2 & 12 \end{pmatrix}.$$

In how many minutes will all the students be in their original places again?

A *permutation* of a set is a list of the elements of this set in some order. More strictly speaking, a *permutation* of a set is a one-to-one mapping of the set onto itself (that is, a bijection).

A permutation f can be conveniently represented as an *oriented graph* whose nodes are elements of a set and whose edges go from node a_k to node $f(a_k)$. A permutation of a set which takes a_k to $f(a_k)$ is written as

$$\begin{pmatrix} a_1 & a_2 & \dots & a_n \\ f(a_1) & f(a_2) & \dots & f(a_n) \end{pmatrix}.$$

Conventionally, $a_k = k$ for $k = 1, \dots, n$.

The *inverse* permutation of the permutation f is the permutation f^{-1} defined by $f(f^{-1}(x)) = x$. It is written as

$$\begin{pmatrix} f(a_1) & f(a_2) & \dots & f(a_n) \\ a_1 & a_2 & \dots & a_n \end{pmatrix}.$$

A *composition* of the permutations f and g is the permutation defined by $(f \circ g)(x) := f(g(x))$.

4.1.2. Find the compositions

(a) $\begin{pmatrix} 1 & 2 & 3 \\ 2 & 1 & 3 \end{pmatrix} \circ \begin{pmatrix} 1 & 2 & 3 \\ 3 & 1 & 2 \end{pmatrix}$;

(b) $\begin{pmatrix} 1 & 2 & 3 \\ 2 & 3 & 1 \end{pmatrix} \circ \begin{pmatrix} 1 & 2 & 3 \\ 3 & 1 & 2 \end{pmatrix}$.

A *cycle* $(a_1, a_2, \ldots, a_n)$ is the permutation

$$\begin{pmatrix} a_1 & a_2 & \ldots & a_{n-1} & a_n \\ a_2 & a_3 & \ldots & a_n & a_1 \end{pmatrix}$$

of a set containing the elements $a_1, a_2, \ldots, a_n$ (and possibly other elements), which takes a_n to a_1 and a_i to a_{i+1} for any $i < n$, and maps each of the other elements of the set to itself.

In this language, the results of problem 4.1.2 can be briefly expressed as follows: $(12) \circ (132) = (13)$ and $(123) \circ (132) = (1)$.

4.1.3. Find the compositions (of permutations on the set of numbers)
(a) $(12) \circ (23)$;   (b) $(23) \circ (12)$;   (c) $(12) \circ (13) \circ (12)$;
(d) $(12345) \circ (12)$;   (e) $(12345) \circ (56789)$.
Give the answers in the form of compositions of disjoint cycles.[1]

Below, we will omit writing $\circ$ to indicate composition.

4.1.4. For any permutation f, there exists an integer $n > 0$ for which $f^n = \mathrm{id}$, that is, after an n-fold application of the permutation, each element goes to itself.

The smallest positive integer n for which $f^n = \mathrm{id}$ is called the *order*, denoted by $\mathrm{ord}\, f$, of the permutation.

4.1.5. Are there any permutations of a 9-element set that have order $7; 10;$ $12; 11$?

---

[1]Compositions of *disjoint cycles* are compositions of cycles that have no common elements, for example, the right-hand side of the equation $(123) \circ (234) = (12) \circ (34)$.

4.1.6. What is the order of the composition of disjoint cycles of lengths $n_1, \ldots, n_k$?

The permutations from problem 4.1.6 of a set of $(n_1 + \ldots + n_k)$ elements are called permutations of *type* $\langle n_1, \ldots, n_k \rangle$. For example, $(14)(253)$ and $(15)(432)$ are of type $\langle 2, 3 \rangle$, and $(1)(3)(245)$ is of type $\langle 1, 1, 3 \rangle$.

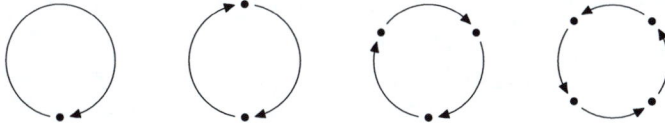

FIGURE 4.1. A permutation of type $\langle 1, 2, 3, 4 \rangle$.

4.1.7. Find the number of permutations of the following types:
(a) $\langle 2, 3 \rangle$; (b) $\langle 3, 3 \rangle$; (c) $\langle 1, 2, 3, 4 \rangle$.

Permutations a and b are called *conjugate* if $a = xbx^{-1}$ for some permutation x.

4.1.8. (a) Permutations a and b are conjugate if and only if their types are the same.

(b) Let a and x be arbitrary permutations of the n-element set. Then

$$xax^{-1} = \begin{pmatrix} x(1) & x(2) & \ldots & x(n) \\ x(a(1)) & x(a(2)) & \ldots & x(a(n)) \end{pmatrix}.$$

In other words, the cyclic decomposition of the permutation xax^{-1} is obtained from the cyclic decomposition of the permutation a by replacing each element with its x-image: if $a = \prod\limits_{j=1}^{q} (i_{j,1}, i_{j,2}, \ldots, i_{j,s_j})$, then

$$xax^{-1} = \prod\limits_{j=1}^{q} (x(i_{j,1}), x(i_{j,2}), \ldots, x(i_{j,s_j})).$$

(c) Find $gf^{-1}g^{-1}f$ for $f := (1, 2, \ldots, N)$ and $g := (N, N+1, \ldots, L)$.

(d) The rotations of a cube around its long diagonals generate conjugate permutations of the set of its vertices.

4.1.9. Any permutation can be represented as a composition of
(a) disjoint cycles;
(b) *transpositions* (cycles of length 2);
(c) transpositions of the form $(1, i), i = 2, 3, \ldots, n$.

4.1.10. Find *two* permutations whose compositions can be used to obtain any permutation of an n-element set.

Hints and answers

4.1.1. *Answer*: After 105 minutes.

4.1.3. *Answer*: (a) (123); (b) (132); (c) (23); (d) (1345); (e) (123456789).

4.1.5. *Answer*: There is no permutation of order 11, but the others exist.

4.1.6. *Answer*: $\mathrm{LCM}(n_1, \ldots, n_k)$.

4.1.7. *Answers*: (a) 20; (b) $4\binom{6}{3}/2 = 40$; (c) $10!/4!$.

4.1.8. (a) *Hint*. Renumber the elements of the set so that the permutation a becomes b. This yields the required permutation x.

(c) *Answer*: $(N-1, N, N+1)$.

4.1.10. *Answer*: For example, (12) and (123 … n).

2. The parity of a permutation (1)

4.2.1. (a) Can an arbitrary permutation be represented as a composition of 3-cycles?

(b) Can an arbitrary permutation be represented as a composition of an even number of transpositions?

(c) **The Russian 15-challenge.** Consider a 4×4 grid containing 15 square pieces of size 1×1 labeled $1, 2, \ldots, 15$, with one open (empty) square. Initially, the squares are arranged as in the figure on the right, with the empty square indicated by *. Is it possible, by sequentially moving the squares to an open square, to change the arrangement to the one shown on the left?

$$
\begin{bmatrix}
1 & 2 & 3 & 4 \\
5 & 6 & 7 & 8 \\
9 & 10 & 11 & 12 \\
13 & 14 & 15 & *
\end{bmatrix}
\qquad
\begin{bmatrix}
1 & 2 & 3 & 4 \\
5 & 6 & 7 & 8 \\
9 & 10 & 11 & 12 \\
13 & 15 & 14 & *
\end{bmatrix}
$$

If you cannot solve problem 4.2.1, continue reading.

Let f be permutation of $\{1, 2, \ldots, n\}$. Call the pair (i, j), $1 \le i, j \le n$, a *disorder* for f if $i < j$ but $f(i) > f(j)$. A permutation is called *even* if the total number of its disorders is even. A permutation is called *odd* if it is not even.

4.2.2. Which cycles $(a_1, \ldots, a_n)$ are even?

4.2.3. (a) The composition of an even (odd) permutation and a transposition is odd (even).

(b) What is an appropriate theorem on the parity of the composition of permutations if we know the parity of each factor?

4.2.4. The following conditions are equivalent.

(a) A permutation can be represented as a composition of an even number of transpositions.

(b) Any representation of a permutation as a composition of transpositions contains an even number of them.

(c) A permutation can be represented as a composition of (possibly zero) 3-cycles.

4.2.5. Let S_n denote the set of all permutations of an n-element set.

(a) In S_n, which set has more elements—the set of even or the set of odd permutations?

(b) For any n and k, find the minimum number of transpositions whose compositions yield an arbitrary permutation of S_n consisting of k disjoint cycles of length greater than 1.

4.2.6.* A permutation x is *generated by the permutations* $p_1, p_2, \ldots, p_k$ if $x = x_1 x_2 \ldots x_n$ and each factor x_i equals some p_j.

(a) Any even permutation is generated by any pair of cycles (each of length ≥ 2) that have exactly one common element and contain all elements of the set.

(b) If nk is even with $n > 1$ and $k > 1$, then the cycles $(1 \ldots n)$ and $(n \ldots n + k - 1)$ generate all permutations in S_{n+k-1}.

(c) If nk is odd with $n > 1$ and $k > 1$, then the cycles $(1 \ldots n)$ and $(n \ldots n + k - 1)$ generate all even permutations (and no others).

Hints and answers

4.2.1. *Answers*: (a) No; (b) No; (c) No.

4.2.2. *Answer*: A cycle of length n is even if n is odd, and is odd if n is even.

4.2.3. *Hint.* Sum the parities of the factors modulo 2.

4.2.5. *Answer*: (a) For $n > 1$ there is an equal number of odd and even permutations; (b) $n - k$.

4.2.6. See I. Grigoriev, *Generation of permutations by figure "eight"*,
`http://www.mccme.ru/mmks/dec10/grigoriev_report.pdf`.
(Be careful—there are flaws!).

3. The combinatorics of equivalence classes (2)

This section is devoted to counting the number of different equivalence classes (such as colorings). These computations will lead to the important notion of a group of transformations and to an elementary formulation of Burnside's lemma. The formulation and proof of this and other results in the abstract language of group theory makes them less accessible.

We do not require that all given colors be used in a coloring. When a coloring is transformed under a rotation, the new coloring is considered the same as the old one (see 4.3.1 (c) for an exception).

The following definitions are used only in 4.3.1 (b), 4.3.6 (e), and 4.3.11, and therefore can be skipped when doing other problems.

An *isomorphism* between two graphs is a bijection between the vertex sets of these graphs such that any two vertices in one graph are connected by an edge if and only if their images under the bijection are connected by an edge in the second graph. An *automorphism* of a graph is an isomorphism of the graph with itself.

4.3.1. (a) How many different ways can one color the faces of a cube in red and gray?

(b) How many different (i.e., nonisomorphic) non-oriented graphs with four vertices are there?

(c) How many different ways can one color the vertices of a regular tetrahedron using r colors?

(Here we view two colorings to be the same if one can be obtained from the other by a not necessarily orientation-preserving motion, for example, a reflection.)

4.3.2. For a prime p, find the number of closed oriented connected length-p circuits (possibly self-intersecting) passing through all the vertices of a given regular p-gon. (The edges of the circuit are sides or diagonals of the regular p-gon. Circuits that coincide after a rotation are considered to be the same; thus, for example, 12543 and 14532 are indistinguishable circuits of a regular 5-gon.)

Problems 4.3.1 and 4.3.2 are simple and can be solved without using Burnside's lemma.

4.3.3. Find the number of colorings in r colors of a circular track ("carousel") consisting of n unlabeled train cars (that is, the number of colorings of the vertices of a regular n-gon in r colors if colorings coinciding after rotation are indistinguishable) for
  (a) $n = 5$;   (b) $n = 4$;   (c) $n = 6$.

For arbitrary n, problem 4.3.3 can be solved by generalizing the methods used for small n, but this can be cumbersome. We give a simple method to deal with trickier values of n by presenting an alternative solution of (c).

Call a coloring of the carousel of *numbered* cars in r colors a *(painted) train*. Clearly, there are a total of r^6 painted 6-car trains.

Distribute trains between stations so that at each station we place all the different trains that can be obtained from a single coloring of the carousel (by decoupling two cars in the carousel). Then the required number Z of colorings is equal to the number of stations.

Let us define the *period* $T(\alpha)$ of the train α to be the smallest positive value of the cyclic shift that takes the train α to itself.

4.3.4. The number of trains at a station is equal to the period of each of the trains standing at this station. In particular, the periods of trains at the same station are equal.

At each station, choose one train and put 6 passengers in it. Give each person a different ticket labeled with one of the numbers $0, 1, 2, 3, 4, 5$. We want to find the total number of passengers, which equals $6Z$.

We instruct each passenger to go to the (painted) train obtained from the chosen one by the cyclic shift indicated by the passenger's ticket number. Clearly every passenger stays at the same station.

4.3.5. (a) In train α there are $6/T(\alpha)$ passengers left. More formally, the number of those s's for which the cyclic shift by s translates the train α into itself is $6/T(\alpha)$.

(b) Each train α will have $6/T(\alpha)$ passengers.

This means that the total number $6Z$ of passengers is equal to the number of all pairs in which $s \in \{0, 1, 2, 3, 4, 5\}$ and α is a train that is unchanged after a cyclic shift of cars. A cyclic shift by s leaves exactly $r^{\gcd(s,6)}$ trains unchanged. Therefore

$$6Z = r^6 + r + r^2 + r^3 + r^2 + r.$$

The above formula is expressed as follows:

$$6Z = \sum_x T(x) \cdot \frac{6}{T(x)} = \sum_\alpha \frac{6}{T(\alpha)} = r^6 + r + r^2 + r^3 + r^2 + r.$$

Here, the first summation is over all the colorings of the carousels, and the second one is over all trains α.

4.3.6. Find the number of colorings of
   (a) a carousel of n cars using r colors;
   (b) necklaces of $n = 2k + 1$ beads using r colors (two necklaces are considered to be the same if one is transformed into the other after rotating around the center of the necklace or after turning the necklace over);
   (c) unnumbered faces of a cube using r colors;
   (d) unnumbered vertices of a cube using r colors;
   (e) unnumbered vertices of the graph $K_{3,3}$ (Figure 4.2) using r colors. Two colorings are considered to be the same if one can be transformed into the other by an automorphism of this graph.

FIGURE 4.2. The graph $K_{3,3}$

4.3.7. List all rotations of the cube, that is, the rotations of space that map the cube to itself.

Here is a plan for attacking problem 4.3.6 (c). Parts (b)–(e) can be solved similarly. Part (b) can be solved even without this hint.

Call the coloring of the *numbered* faces of the cube using r colors the *(colored) box* (or *frozen coloring*). Then there are a total of r^6 boxes. Place the boxes in rooms so that each room contains all the boxes obtained from some box by various rotations. Hence the number of distinct colorings, Z, is equal to the number of rooms.

In each room, choose one box and put 24 cockroaches in it, corresponding to the rotations of the cube. We need to count the total number of cockroaches, which equals $24Z$.

Instruct each cockroach to crawl into the box obtained from the chosen one by the rotation that corresponds to that cockroach. It is clear that every cockroach stays in the same room.

The number of cockroaches remaining in the chosen box is equal to the number of cube rotations that turn this box into itself. Let $\mathrm{st}(\alpha)$ denote the number of cube rotations that leave the (painted) box (that is, the frozen coloring) α unchanged.

4.3.8. (a) The number of cockroaches in box α is equal to $\text{st}(\alpha)$. More formally, if there is a rotation that turns the frozen coloring α into the frozen coloring α', then the number of such rotations is equal to $\text{st}(\alpha)$.

(b) In any other box from the selected room, there will be as many cockroaches as in the chosen box in the same room. More formally, for any two frozen colorings α and α' that turn into each other by some rotation, we have $\text{st}(\alpha) = \text{st}(\alpha')$. This number is denoted by $\text{st}(x)$, where x is the corresponding coloring of unnumbered faces of the cube.

Therefore, the total number of cockroaches is equal to the number of pairs (α, s) for which s is a rotation of the cube and α is a box left unchanged by s. So it remains to solve the following problem.

4.3.9. For each rotation s of the cube find the number of boxes (frozen colorings) left unchanged under s.

Denote by N_x the number of frozen colorings corresponding to the coloring of x. Then for any coloring x, the number $\text{st}(x) \cdot N_x$ is equal to the number of rotations of the cube (i.e., to 24). In other words,

$$24Z = \sum_x \text{st}(x) \cdot N_x = \sum_\alpha \text{st}(\alpha) = \sum_s \text{fix}(s).$$

Here, the first summation is over all colorings x of unnumbered faces, the second is over all frozen colorings α, and the third is over all rotations s of the cube.

Can we formulate a general result that could be applied instead of repeating the arguments of problem 4.3.6 (a) and (c)?

4.3.10. Burnside's lemma. Let M be a set and let $G = \{g_1, g_2, \ldots, g_n\}$ be a family of transformations of this set that is closed with respect to composition and inverse.[2] Two elements of the set M are called *equivalent* if one of them can be transformed into the other by one of these transformations. Then the number of equivalence classes is equal to $\frac{1}{n} \sum_{k=1}^{n} \text{fix}(g_k)$, where $\text{fix}(g_k)$ is the number of elements of M that are left unchanged ("fixed") by g_k.

4.3.11. Find the number of graphs with n vertices, up to isomorphism. The answer can be left as a sum.

---

[2] In other words, G is a *group* of transformations of M.

4.3.12. (a) Find the number b_n of mappings $\{0,1\}^n \to \{0,1\}$ up to variable permutations.

(b) Prove that there exists a limit $\lim_{n \to \infty} n! b_n / 2^{2^n}$ and find it. (For the definition of *limit*, see problem 6.4.2. Skip this problem if you are unfamiliar with limits.)

Answers

4.3.1. *Answers*: (a) 10; (b) 11; (c) $r(r+1)(r+2)(r+3)/24$.

4.3.2. *Answer*: $p - 2 + ((p-1)! + 1)/p$.

4.3.3. *Answers*: (a) $(r^5 + 4r)/5$; (b) $(r^4 + r^2 + 2r)/4$; (c) $(r^6 + r^3 + 2r^2 + 2r)/6$.

4.3.6. (a) *Answer*: $\frac{1}{n} \sum_{d \mid n} \varphi\left(\frac{n}{d}\right) r^d$. Euler's function $\varphi(n)$ is defined in problem 2.1.5.

Chapter 5

Inequalities

This chapter is almost independent of the rest of the book. Only simple facts from it are used in other chapters.

Unless otherwise stated, Roman and Greek letters denote non-negative real numbers. In problem statements, denominators are assumed to be nonzero. With problems involving x^a it is useful to first consider rational a; you can stop at this if you do not know what $2^{\sqrt{2}}$ is. After proving a non-strict inequality, it is useful to consider how and under what conditions it can turn into an equality. In this case, it is also useful to check that all intermediate inequalities used in the proof of the original one turn into equalities.

1. Towards Jensen's inequality (2)

Remember to prove all the inequalities you use!

5.1.1. Paul took a physics and mathematics olympiad lasting 6 hours. He receives x and y points (not necessarily integers) for the time he spends on physics and mathematics problems respectively. How should he distribute time between physics and mathematics in order to obtain the highest (lowest) total result if this result is obtained by the formula

(a) xy;   (b) $x^2 + y^2$;   (c) $\sqrt{x} + \sqrt{y}$;   (d) $\frac{1}{x} + \frac{1}{y}$;

(e) $\sin \frac{x}{2} + \sin \frac{y}{2}$;   (f) $x^2 y$.

Let $I \subset \mathbb{R}$ be a finite or infinite interval. A function $f \colon I \to \mathbb{R}$ is said to be *concave up* if

$$f\left(\frac{x+y}{2}\right) \leq \frac{f(x) + f(y)}{2} \quad \text{for any } x, y \in I.$$

5.1.2. Solve problem 5.1.1 for the formula $f(x) + f(y)$, where the function f is concave up.

5.1.3. Which of these functions are concave up and over what intervals?
(a) x;   (b) x^2;   (c) $-x^2$;   (d) $(x-1)^3$;   (e) $\sqrt{x}$;   (f) $|x-3|$?

5.1.4. Let $n > 0$ be an integer and let $x_1 + \ldots + x_n = 1$. Find the highest and the lowest values of the following expressions:
(a) $x_1 \cdot \ldots \cdot x_n$;
(b) $x_1^2 + \ldots + x_n^2$;
(c) $\frac{1}{x_1} + \ldots + \frac{1}{x_n}$;
(d) $x_1^3 + \ldots + x_n^3$;
(e)* $x_1^2 \cdot \ldots \cdot x_{n-1}^2 \cdot x_n$.

5.1.5. Inequalities for mean values, or Cauchy's inequalities. Prove the following inequalities:

$$\min\{x_1, \ldots, x_n\} \leq \frac{n}{\frac{1}{x_1} + \ldots + \frac{1}{x_n}} \leq \sqrt[n]{x_1 \cdot \ldots \cdot x_n}$$

$$\leq \frac{x_1 + \ldots + x_n}{n} \leq \sqrt{\frac{x_1^2 + \ldots + x_n^2}{n}} \leq \max\{x_1, \ldots, x_n\}.$$

5.1.6. Solve problem 5.1.1 (b–e) in the case where Paul solves physics problems twice as fast as mathematics problems but the solution of a math problem has twice the value of a solved physics problem. More formally, find the largest and the smallest values of the expression $2f(x) + f(2y)$ provided that $x + y = 6$ for $f(x) = x^2, \sqrt{x}, \frac{1}{x}, \sin \frac{x}{2}$.

5.1.7.* (a) If a continuous function f is concave up then

$$f(tx + (1-t)y) \leq tf(x) + (1-t)f(y) \quad \text{for any } t \in [0,1] \text{ and } x, y \in I.$$

(b) **Jensen's inequality.** If a continuous function $f \colon I \to \mathbb{R}$ is concave up then

$$f(t_1 x_1 + \ldots + t_n x_n) \leq t_1 f(x_1) + \ldots + t_n f(x_n)$$

for any $t_1, \ldots, t_n$ whose sum is 1 and any $x_1, \ldots, x_n \in I$.[1]

(c) *We can check whether a function is concave up using the second derivative.* A function with continuous second derivative is concave up if and only if $f''(x) \geq 0$ on I. For a proof, we need analogues of some results in section 2 of Chapter 7 for differentiable functions.

---

[1] One can define concave down functions analogously, and then Jensen's inequality holds with the inequality reversed. See the alternative solution to problem 5.2.5 (b) on p. 75.

5.1.8.* (a) Find the largest and smallest values of $x_1 \cdot \ldots \cdot x_n$ under the conditions $x_i \geq \frac{1}{n}$ and $x_1^2 + \ldots + x_n^2 = 1$.

(b) Prove the inequality

$$x_1 x_2 + x_2 x_3 + \ldots + x_{n-1} x_n \leq \frac{(x_1 + \ldots + x_n)^2}{4}.$$

5.1.9.* A set of numbers $x_1 \geq \ldots \geq x_n$ from I *majorizes* a set of numbers $y_1 \geq \ldots \geq y_n$ from I if $x_1 + \ldots + x_n = y_1 + \ldots + y_n$ and $x_1 + \ldots + x_k \geq y_1 + \ldots + y_k$ for each k.

(a) **Karamata's inequality.** If a continuous function $f \colon I \to R$ is convex down and a set of numbers $x_1 \geq \ldots \geq x_n$ from I majorizes a set of numbers $y_1 \geq \ldots \geq y_n$ from I, then $f(x_1) + \ldots + f(x_n) \geq f(y_1) + \ldots + f(y_n)$.

(b) Prove the inequality

$$\sqrt{a+b-c} + \sqrt{b+c-a} + \sqrt{c+a-b} \leq \sqrt{a} + \sqrt{b} + \sqrt{c}$$

for sides a, b, c of a triangle.

(c) Find the smallest C such that for any $x_1, \ldots, x_9 \geq 0$ the following inequality holds:

$$\sum_{1 \leq i < k \leq 9} x_i x_k (x_i^2 + x_k^2) \leq C \left(\sum_{i=1}^{9} x_i \right)^4.$$

Hints

5.1.4. (a) *Hint.* Below we describe a common mistake along with a useful technique that corrects it.

Suppose you want to prove that the product $x_1 \cdot \ldots \cdot x_n$ attains a maximum value of $\left(\frac{1}{n} \right)^n$ with $x_1 = x_2 = \ldots = x_n = \frac{1}{n}$. If among the numbers $x_1, x_2, \ldots, x_n$ satisfying the condition $x_1 + \ldots + x_n = 1$ there is a pair of distinct numbers x_i and x_j, then replace them by $\frac{x_i + x_j}{2}$ and $\frac{x_i + x_j}{2}$. The sum of the numbers of the new set remains 1, and the product will become strictly larger, since $x_i x_j < \left(\frac{x_i + x_j}{2} \right)^2$. Hence, to maximize the product, the values of x_i must satisfy $x_1 = x_2 = \ldots = x_n$.

These arguments are not enough to solve the problem. There is no proof that there is a set of $x_1, x_2, \ldots, x_n$ for which the product $x_1 \cdot \ldots \cdot x_n$ attains a greatest value. This can be accomplished using results about the existence of extrema for polynomials of several variables, that is, an analogue of Theorem 7.2.7 (a). However, it is simpler to change the argument as follows.

Let $x_1, x_2, \ldots, x_n$ be non-negative numbers whose sum is 1. If not all of them are equal to $\frac{1}{n}$, then there are numbers $x_i < \frac{1}{n}$ and $x_j > \frac{1}{n}$. Replace the numbers x_i and x_j with $\frac{1}{n}$ and $x_i + x_j - \frac{1}{n}$. The difference between the numbers of the new pair (with the same sum) becomes smaller and therefore their product increases (check this!). Hence the product of numbers in the new set also increases. Therefore, making no more than n replacements, we get the set $\frac{1}{n}, \frac{1}{n}, \ldots, \frac{1}{n}$. The product of the numbers in each next set is greater than the previous one. Therefore $x_1 \cdot \ldots \cdot x_n \leq \frac{1}{n} \cdot \ldots \cdot \frac{1}{n} = \left(\frac{1}{n}\right)^n$.

Suggestions, solutions, and answers

5.1.1. (a) *Answer*: The maximum is attained when $x = y = 3$, and the minimum, for example, when $x = 0$ and $y = 6$.

Solution. It is easy to see that the minimum value of xy is zero for $x \geq 0$ and $y \geq 0$. Since

$$xy = \frac{(x+y)^2 - (x-y)^2}{4} = 9 - \frac{(x-y)^2}{4},$$

the maximum value of xy is attained when the variables take equal values.

(b) *Answer*: The minimum is attained when $x = y = 3$, and the maximum, for example, when $x = 0$ and $y = 6$.

Hint. We have $x^2 + y^2 = \frac{(x+y)^2 + (x-y)^2}{2} = 18 + \frac{(x-y)^2}{2}$.

(c) *Answer*: The maximum is attained when $x = y = 3$, and the minimum, for example, at $x = 0$ and $y = 6$.

Solution. Since $(\sqrt{x} + \sqrt{y})^2 = x + y + 2\sqrt{xy} = 6 + 2\sqrt{xy}$, the expression $\sqrt{x} + \sqrt{y}$ attains its maximum (minimum) value simultaneously with xy. Thus, the answer is the same as in part (a).

(d) *Answer*: The minimum is attained at $x = y = 3$, and there is no maximum.

Solution. We have $\frac{1}{x} + \frac{1}{y} = \frac{x+y}{xy} = \frac{6}{xy}$. Therefore, according to part (a), the minimum is attained when $x = y = 3$,

Since $\frac{1}{1/a} + \frac{1}{6-1/a} > a$ for any $a > 0$, the value of $\frac{1}{x} + \frac{1}{y}$ can be made arbitrarily large.

(e) *Answer*: The maximum is attained when $x = y = 3$, and the minimum, for example, at $x = 0$ and $y = 6$.

Solution. We have

$$\sin\frac{x}{2} + \sin\frac{y}{2} = 2\sin\frac{x+y}{4}\cos\frac{x-y}{4} = 2\sin\frac{3}{2}\cos\frac{x-y}{4}.$$

Since $\frac{3}{2} < \pi$, we have $\sin\frac{3}{2} > 0$. Since $\cos x \leq 1$, the maximum is attained when $x = y = 3$.

The function $\cos x$ decreases on the interval $\left[0, \frac{\pi}{2}\right]$ and increases on $\left[-\frac{\pi}{2}, 0\right]$. We have $\frac{|x-y|}{4} \leq \frac{3}{2} < \frac{\pi}{2}$. Therefore the minimum is attained when $x - y = \pm 6$.

5.1.2. (a) For the maximum, Paul needs to solve all the problems in one subject exclusively, and for the minimum, he needs to distribute time equally between the subjects.

5.1.4. *Answers.*
(a) The maximum is equal to $(1/n)^n$ and the minimum is equal to 0.
(b) The maximum is equal to 1 and the minimum is equal to $1/n$.
(c) The maximum does not exist and the minimum is equal to n^2.
(d) The maximum is equal to 1 and the minimum is equal to $1/n^2$.

5.1.6. *Answers.*
For the function x^2 the minimum is attained when $x = 2y = 4$, and the maximum when $(x, y) = (0, 6)$.
For the function $\sqrt{x}$, the maximum is attained when $x = 2y = 4$, and the minimum when $(x, y) = (6, 0)$.
For the function $1/x$, the minimum is attained when $x = 2y = 4$, and there is no maximum.
For the function $\sin \frac{x}{2}$ everything is more complicated. This is a special case of the inequality in 5.1.7 (a).

5.1.7. *Hints.*
(a) First prove the statement for binary rational t. For arbitrary m, take a limit.
(b) Deduce the statements from (a) using induction on n.
(c) Use Lagrange's Mean Value Theorem 7.2.7 (c).

2. Some basic inequalities (2)

5.2.1. (a) If $a \geq b$ and $x \geq y$, then $ax + by \geq ay + bx$.
(b) If a, b, c and α, β, γ are the sides and angles of a triangle, respectively (with angle α opposite to side a, etc), then

$$a\alpha + b\beta + c\gamma \geq \frac{\pi}{3}(a + b + c).$$

(c) If α, β, and γ are the angles of a triangle, then

$$2\left(\frac{\sin \alpha}{\alpha} + \frac{\sin \beta}{\beta} + \frac{\sin \gamma}{\gamma}\right) \leq \left(\frac{1}{\alpha} + \frac{1}{\beta}\right)\sin \gamma + \left(\frac{1}{\alpha} + \frac{1}{\gamma}\right)\sin \beta + \left(\frac{1}{\beta} + \frac{1}{\gamma}\right)\sin \alpha.$$

(d) **Rearrangement inequality.** If $x_1 \geq \ldots \geq x_n$, $y_1 \geq \ldots \geq y_n$, and $\{i_1, \ldots, i_n\}$ is any permutation of $\{1, \ldots, n\}$, then

$$x_1 y_1 + \ldots + x_n y_n \geq x_1 y_{i_1} + \ldots + x_n y_{i_n} \geq x_1 y_n + \ldots + x_n y_1.$$

(e) **Chebyshev's inequality.** If $x_1 \geq \ldots \geq x_n$ and $y_1 \geq \ldots \geq y_n$, then

$$\frac{x_1 y_1 + \ldots + x_n y_n}{n} \geq \frac{x_1 + \ldots + x_n}{n} \cdot \frac{y_1 + \ldots + y_n}{n} \geq \frac{x_1 y_n + \ldots + x_n y_1}{n}.$$

5.2.2. Prove the following inequalities.
(a) $a^3 + b^3 \geq a^2 b + ab^2$;
(b) $a^k + b^k \geq a^{k-l}b^l + a^l b^{k-l}$ for any $k > l > 0$;
(c) $a^2 + b^2 + c^2 \geq ab + bc + ca$;
(d) $2(a^3 + b^3 + c^3) \geq a^2 b + b^2 a + b^2 c + c^2 b + c^2 a + a^2 c \geq 6abc$;
(e) $2(a^4 + b^4 + c^4) \geq a^3 b + b^3 a + b^3 c + c^3 b + c^3 a + a^3 c$
$$\geq 2(a^2 b^2 + b^2 c^2 + c^2 a^2) \geq 2abc(a + b + c).$$

5.2.3. (Challenge.) Find and prove a chain of inequalities similar to the previous problem: (a) between $a^5 + b^5 + c^5$ and $abc(ab + bc + ca)$;
(b) between $a^k + b^k + c^k$ and $a^q b^q c^q M$, where $k > q \geq 0$ are integers, $x^0 := 1$ for $x \in \{a, b, c\}$, and

$$M = \begin{cases} 1, & n = 3q, \\ a + b + c, & n = 3q + 1, \\ ab + bc + ca, & n = 3q + 2; \end{cases}$$

(c) starting with $a_1^k + a_2^k + \ldots + a_n^k$, where $k > 0$ is an integer.

5.2.4. (a) If $at^2 + 2bt + c \geq 0$ for any t, then $b^2 \leq ac$.
(b) **Cauchy–Buniakovsky–Schwarz (CBS) inequality.** Prove the following inequality:

$$(a_1 b_1 + \ldots + a_n b_n)^2 \leq (a_1^2 + \ldots + a_n^2)(b_1^2 + \ldots + b_n^2).$$

A geometric interpretation (which we will not use below) is that the scalar product of two vectors in an n-dimensional space does not exceed the product of their lengths.
(c) Equality in part (b) is achieved only with *proportional* sequences $a_1, a_2, \ldots, a_n$ and $b_1, b_2, \ldots, b_n$, i.e., sequences such that $a_1/b_1 = a_2/b_2 = \ldots = a_n/b_n$.
(d) The following inequality holds:

$$\frac{x_1^2}{y_1} + \frac{x_2^2}{y_2} + \ldots + \frac{x_n^2}{y_n} \geq \frac{(x_1 + x_2 + \ldots + x_n)^2}{y_1 + y_2 + \ldots + y_n}.$$

5.2.5. Prove the inequalities below.
(a) $x^3 + 2y^{3/2} \geq 3xy$.
(b) **Young's inequality.** If $\frac{1}{p} + \frac{1}{q} = 1$, then $xy \leq \frac{x^p}{p} + \frac{y^q}{q}$.
(c) **Local inequality.** For any $k > l > 0$, we have $\frac{a^k}{b^l} \geq \frac{ka^{k-l} - lb^{k-l}}{k-l}$.
Equality is achieved only when $a = b$.
(d) **Hölder's inequality.** If $\frac{1}{p} + \frac{1}{q} = 1$, then

$$x_1 y_1 + \ldots + x_n y_n \leq (x_1^p + \ldots + x_n^p)^{\frac{1}{p}} (y_1^q + \ldots + y_n^q)^{\frac{1}{q}}.$$

(e) **Minkowski's inequality.** If $p > 1$, then

$$\left(\sum_{i=1}^{n} (x_i + y_i)^p \right)^{\frac{1}{p}} \leq \left(\sum_{i=1}^{n} x_i^p \right)^{\frac{1}{p}} + \left(\sum_{i=1}^{n} y_i^p \right)^{\frac{1}{p}}.$$

Hints

5.2.3. (c) **Muirhead's inequality.** If a set of numbers $a_1 \geq \ldots \geq a_n \geq 0$ majorizes a set of numbers $b_1 \geq \ldots \geq b_n \geq 0$ (see definition in problem 5.1.9), then

$$\sum_{\sigma \in S_n} x_1^{a_{\sigma(1)}} \cdot \ldots \cdot x_n^{a_{\sigma(n)}} \geq \sum_{\sigma \in S_n} x_1^{b_{\sigma(1)}} \cdot \ldots \cdot x_n^{b_{\sigma(n)}}$$

for any $x_1, \ldots, x_n$. Here $x_s^0 := 1$. A proof can be found, for example, in the article [**DY85**].

Suggestions, solutions, and answers

5.2.4. (b, d) Use induction on n or part (a).

Inequality (d) is equivalent to inequality (b) for the sequences $\frac{x_1}{\sqrt{y_1}}$, $\frac{x_2}{\sqrt{y_2}}, \ldots, \frac{x_n}{\sqrt{y_n}}$ and $\sqrt{y_1}, \sqrt{y_2}, \ldots, \sqrt{y_n}$.

5.2.5. (b) Draw the graph of $y(x) = x^{p-1}$ on the coordinate plane. Shade the region between the graph and the x-axis for $0 \leq x \leq a$. The same curve is the graph of the function $x(y) = y^{q-1}$. Shade the region between it and the y-axis for $0 \leq y \leq b$. The total area of the shaded regions is equal to $\frac{a^p}{p} + \frac{b^q}{q}$. This region contains a rectangle with sides a and b, yielding the inequality.

Alternative solution. Apply Jensen's inequality to the function $f(x) = \ln x$, which is concave down for $x > 0$, with coefficients $\frac{1}{p}$ and $\frac{1}{q}$ and numbers a^p and b^q:

$$\ln \left(\frac{a^p}{p} + \frac{b^q}{q} \right) \geq \frac{1}{p} \ln a^p + \frac{1}{q} \ln b^q = \ln(ab).$$

(c) The inequality follows from part (b).

(d) It suffices to prove the inequality for the case where $x_1^p + \ldots + x_n^p = y_1^q + \ldots + y_n^q = 1$, which follows from Young's inequality.

(e) Apply Hölder's inequality to the right-hand side of the equality below:

$$\sum_{i=1}^{n} (x_i + y_i)^p = \sum_{i=1}^{n} x_i (x_i + y_i)^{p-1} + \sum_{i=1}^{n} y_i (x_i + y_i)^{p-1}.$$

3. Applications of basic inequalities (3*)
By M. A. Bershtein

The author is grateful to A. Bershtein, A. Dudko, V. Karajko, K. Knop, and V. Frank, who taught him almost everything that is written here.

5.3.1. For positive integers a, b, and c, the following inequality holds

$$\left(\frac{a^2 + b^2 + c^2}{a + b + c}\right)^{a+b+c} \geq a^a b^b c^c \geq \left(\frac{a + b + c}{3}\right)^{a+b+c}.$$

5.3.2. (a) **Weighted Cauchy's inequality.** If $a_1 > 0, \ldots, a_n > 0$ and $a_1 + \ldots + a_n = 1$, then $a_1 x_1 + \ldots + a_n x_n \geq x_1^{a_1} \cdot \ldots \cdot x_n^{a_n}$.

(This is a generalization of Young's inequality 5.2.5 (b).)

(b)* Define the *weighted power mean with exponent m of numbers* $x_1, \ldots, x_n$ *with weights* $a_1, \ldots, a_n > 0$, where $a_1 + \ldots + a_n = 1$, by

$$S_m := \sqrt[m]{a_1 x_1^m + \ldots + a_n x_n^m} \quad \text{for } m \neq 0, \quad S_0 := x_1^{a_1} \cdot \ldots \cdot x_n^{a_n},$$
$$S_{-\infty} := \min\{x_1, \ldots, x_n\}, \quad \text{and} \quad S_{+\infty} := \max\{x_1, \ldots, x_n\}.$$

Prove that $S_a \leq S_b$ if $a \leq b$ for any $a, b \in \mathbb{R} \cup \{-\infty, +\infty\}$.

(c) Is it true that if $a \leq b$ then $S_a \leq S_b$ for any positive values of $x_1, \ldots, x_n$ provided that one of a_i's is negative?

5.3.3. Prove the following inequalities.

(a) $\frac{a_1^2}{a_2} + \frac{a_2^2}{a_3} + \ldots + \frac{a_n^2}{a_1} \geq a_1 + a_2 + \ldots + a_n$;

(b) $\frac{a_1^2}{a_1+a_2} + \frac{a_2^2}{a_2+a_3} + \ldots + \frac{a_n^2}{a_n+a_1} \geq \frac{1}{2}(a_1 + a_2 + \ldots + a_n)$.

5.3.4. Prove

$$\frac{a^2}{b(a + c)} + \frac{b^2}{c(b + d)} + \frac{c^2}{d(c + a)} + \frac{d^2}{a(d + b)} \geq 2.$$

5.3.5. Prove the following inequalities.

(a) $a^3 b + b^3 c + c^3 a \geq abc(a + b + c)$;

(b) $a^3 b^2 + b^3 c^2 + c^3 a^2 \geq abc(ab + bc + ca)$.

5.3.6. Prove the following inequalities.

(a) $\frac{a_1^3}{a_1+a_2} + \frac{a_2^3}{a_2+a_3} + \ldots + \frac{a_n^3}{a_n+a_1} \geq \frac{1}{2}(a_1^2 + a_2^2 + \ldots + a_n^2)$;

(b) $\frac{a}{b+2c+d} + \frac{b}{c+2d+a} + \frac{c}{d+2a+b} + \frac{d}{a+2b+c} \geq 1$.

5.3.7. Prove the following inequalities.

(a) $\frac{a}{b+c} + \frac{b}{c+d} + \frac{c}{d+a} + \frac{d}{a+b} \geq 2$;

(b) $\frac{a+c}{a+b} + \frac{b+d}{b+c} + \frac{c+a}{c+d} + \frac{d+b}{d+a} \geq 4$.

5.3.8. Prove that if $ab + bc + cd + da = 1$, then

$$\frac{a^3}{b+c+d} + \frac{b^3}{c+d+a} + \frac{c^3}{d+a+b} + \frac{d^3}{a+b+c} \geq \frac{1}{3}.$$

5.3.9.* If P, S, and $\alpha_1, \ldots, \alpha_n$ are the perimeter, area, and angles of a convex n-gon, respectively, then

$$P^2 \geq 4S \sum_{i=1}^{n} \cot \frac{\alpha_i}{2}.$$

5.3.10. (a) If A', B', and C' are the points where the angle bisectors of the triangle ABC intersect the opposite sides and I is the center of the inscribed circle, then $\frac{1}{4} < \frac{AI \cdot BI \cdot CI}{AA' \cdot BB' \cdot CC'} \leq \frac{8}{27}$.

(*International Mathematical Olympiad*, 1994.)

(b) Prove that

$$a^4 + b^4 + c^4 + d^4 + 2abcd \geq a^2b^2 + a^2c^2 + a^2d^2 + b^2c^2 + b^2d^2 + c^2d^2.$$

(c) Prove that

$$x^6y^6 + y^6z^6 + z^6x^6 + 3x^4y^4z^4 \geq 2(x^3 + y^3 + z^3)x^3y^3z^3.$$

5.3.11.* Let a, b, and c be positive numbers with product equal to 1.

(a) Then $\frac{1}{a^3(b+c)} + \frac{1}{b^3(a+c)} + \frac{1}{c^3(a+b)} \geq \frac{3}{2}$.

(*International Mathematical Olympiad*, 1995.)

(b) Find all $\alpha \in \mathbb{R}$ for which the following inequality holds:

$$\frac{a^\alpha}{b+c} + \frac{b^\alpha}{a+c} + \frac{c^\alpha}{a+b} \geq \frac{3}{2}.$$

Hints

An inequality of the form $P(x_1, x_2, \ldots, x_n) \geq 0$ is said to be *symmetric* if $P(x_1, x_2, \ldots, x_n)$ is invariant under any permutation of the variables $x_1, x_2, \ldots, x_n$. An inequality of the form $P(x_1, x_2, \ldots, x_n) \geq 0$ is called *cyclic* if $P(x_1, x_2, \ldots, x_n)$ is invariant under a cyclic permutation of the variables (which takes x_1 to x_2, x_2 to x_3, x_3 to $x_4, \ldots$, and x_n to x_1). Muirhead's inequality 5.2.3 (c), the CBS inequality 5.2.4 (b, d), and Young's inequality 5.2.5 (c) are very useful for proving symmetric and cyclic inequalities. It is easier to apply the CBS inequality in the form 5.2.4 (d) (for example, in inequalities 5.3.3, 5.3.4, and 5.3.6), as you can avoid complex substitution with radicals (see the second solution to problem 5.3.6 (b) below; in other cases you can easily make the necessary substitutions yourself). Similarly, Young's inequality is often easier to apply in the form of 5.2.5 (c).

Suggestions, solutions, and answers

In the solutions of problems 5.3.1, 5.3.7, and 5.3.8 we used K. Oganesyan's work [**Siv67**, problem 204].

5.3.1. From the inequality 5.1.5 for mean values we have

$$\frac{(\underbrace{a+a+\ldots+a}_{a \text{ times}}) + (\underbrace{b+b+\ldots+b}_{b \text{ times}}) + (\underbrace{c+c+\ldots+c}_{c \text{ times}})}{a+b+c}$$

$$\geq \frac{a+b+c}{(\underbrace{\frac{1}{a}+\frac{1}{a}+\ldots+\frac{1}{a}}_{a \text{ times}}) + (\underbrace{\frac{1}{b}+\frac{1}{b}+\ldots+\frac{1}{b}}_{b \text{ times}}) + (\underbrace{\frac{1}{c}+\frac{1}{c}+\ldots+\frac{1}{c}}_{c \text{ times}})}$$

$$= \frac{a+b+c}{3}.$$

The right-hand side of the desired inequality follows by raising both sides to the $a+b+c$ power.

5.3.2. (a) For positive integers $a_1, \ldots, a_n$, the inequality follows from the usual Cauchy inequality 5.1.5, similarly to problem 5.3.1 (here we relax the condition that the a_i's sum to 1). The case of integer values reduces to the case of positive integers, the case of rational values reduces to the case of integers, and the general case reduces to the case of rational values by passing to the limit.

(b) First let $a_1 = \ldots = a_n$. With rational a and b, the proof is similar to the hint for problem 5.1.4. The case of arbitrary a and b is obtained by passing to the limit.

Next, prove the inequality for rational $a_1, \ldots, a_n$. Finally, for arbitrary $a_1, \ldots, a_n$, pass to the limit.

5.3.3. (a) This simple cyclic inequality has several different proofs:

A "global" method is to estimate the entire sum as a whole, applying the CBS inequality 5.2.4 (d).

A "local" method proceeds by estimating each term in the sum, using the inequality $\frac{a^2}{b} \geq 2a - b$ (a special case of Young's inequality in the form of 5.2.5 (c)).

The inequality can be easily proved using the rearrangement inequality 5.2.1 (d).

A method using the weighted Cauchy's inequality 5.3.2 (a) is described in the instructions to problem 5.3.5 (a).

(b) *First method.* We can prove this inequality by estimating each term of the sum on the left-hand side separately, using the inequality $\frac{a^2}{b} \geq 2a - b$, which implies

$$(1) \qquad\qquad \frac{(2a)^2}{a+b} \geq 4a - (a+b).$$

Multiplying the inequality that we wish to prove by 4 yields

$$\frac{4a_1^2}{a_1 + a_2} + \frac{4a_2^2}{a_2 + a_3} + \ldots + \frac{4a_n^2}{a_n + a_1} \geq 2(a_1 + a_2 + \ldots + a_n).$$

Using (1), each term on the left-hand side can be estimated as follows:

$$\frac{(2a_k)^2}{a_k + a_{k+1}} \geq 4a_k - (a_k + a_{k+1}).$$

By adding these, we obtain the required inequality.

Second method. It is possible to prove this inequality by estimating the entire sum on the left-hand side using the CBS inequality 5.2.4 (d).

5.3.5. (a) Divide both sides by abc and reduce the inequality to 5.3.3 (a).

Another way is to use the weighted Cauchy's inequality 5.3.2 (a). Namely, select x, y, and z such that the inequality $xa^3b + yb^3c + zc^3a \geq (x + y + z)a^2bc$ holds. For this it is enough that the following equality holds: $(a^3b)^x(b^3c)^y(c^3a)^z = (a^2bc)^{x+y+z}$. We get the system of linear equations

$$3x + z = 2(x + y + z), \quad 3y + x = x + y + z, \quad 3z + y = x + y + z,$$

which has the solution $x = 4, y = 2, z = 1$. Then we have the inequality $4a^3b + 2b^3c + c^3a \geq 7a^2bc$. Similarly,

$$a^3b + 4b^3c + 2c^3a \geq 7ab^2c \quad \text{and} \quad 2a^3b + b^3c + 4c^3a \geq 7abc^2.$$

Adding these three inequalities, we get

$$a^3b + b^3c + c^3a \geq a^2bc + ab^2c + abc^2.$$

(b) The proof is similar to that of (a). Use the weighted Cauchy's inequality 5.3.2 (a) to prove that

$$4a^3b^2 + 2b^3c^2 + c^3a^2 \geq 7a^2b^2c.$$

Remark 5.3.12. From Muirhead's inequality in the hint for problem (5.2.3) (c), it follows that

$$a^3b + a^3c + b^3a + b^3c + c^3a + c^3b \geq 2a^2bc + 2ab^2c + 2abc^2.$$

Inequality 5.3.5 (a) shows that stronger inequalities are true:

$$a^3b + b^3c + c^3a \geq a^2bc + ab^2c + abc^2 \quad \text{and} \quad a^3c + c^3b + b^3c \geq a^2bc + ab^2c + abc^2.$$

Similarly, inequality 5.3.5 (b) strengthens the following particular case of Muirhead's inequality:

$$a^3b^2 + a^3c^2 + b^3a^2 + b^3c^2 + c^3a^2 + c^3b^2 \geq 2a^2b^2c + 2a^2bc^2 + 2ab^2c^2.$$

5.3.6. *First method.* Estimate each term using the inequality $\frac{a^3}{b} \geq \frac{3a^2 - b^2}{2}$ (a special case of Young's inequality 5.2.5 (c)). Equality is achieved when the cubic root of the numerator in the left-hand side is equal to the left-hand side's denominator. In the original inequality, equality is achieved as usual with $a_1 = a_2 = \ldots = a_n$. Multiply the entire inequality by 8 so that when $a_1 = a_2 = \ldots = a_n$ the cubic root of the numerator equals the denominator. After that, estimate each term on the left-hand side using the inequality $\frac{(2a_1)^3}{a_1 + a_2} \geq \frac{12a_1^2 - (a_1 + a_2)^2}{2}$. This reduces to proving

$$a_1^2 + a_2^2 + \ldots + a_n^2 \geq a_1 a_2 + a_2 a_3 + \ldots + a_n a_1,$$

which is equivalent to

$$(a_1 - a_2)^2 + (a_2 - a_3)^2 + \ldots + (a_n - a_1)^2 \geq 0.$$

Second method. Estimate the total sum on the left-hand side:

$$\frac{a_1^4}{a_1(a_1 + a_2)} + \frac{a_2^4}{a_2(a_2 + a_3)} + \ldots + \frac{a_n^4}{a_n(a_n + a_1)}$$
$$\geq \frac{(a_1^2 + a_2^2 + \ldots + a_n^2)^2}{a_1(a_1 + a_2) + a_2(a_2 + a_3) + \ldots + a_n(a_n + a_1)} \geq \frac{1}{2}(a_1^2 + a_2^2 + \ldots + a_n^2).$$

The first inequality follows from the CBS inequality (5.2.4 (d)), and the second is equivalent to

$$a_1^2 + a_2^2 + \ldots + a_n^2 \geq a_1 a_2 + a_2 a_3 + \ldots + a_n a_1$$

(see the end of the first method).

(b) Use the inequalities

$$\frac{a^2}{a(b + 2c + d)} + \frac{b^2}{b(c + 2d + a)} + \frac{c^2}{c(d + 2a + b)} + \frac{d^2}{d(a + 2b + c)}$$
$$\geq \frac{(a + b + c + d)^2}{a(b + 2c + d) + b(c + 2d + a) + c(d + 2a + b) + d(a + 2b + c)} \geq 1.$$

The first inequality follows from the CBS inequality, and the second one is equivalent to

$$a^2 + b^2 + c^2 + d^2 \geq 2ac + 2bd.$$

Another solution: Apply the CBS inequality to the sequences

$$\sqrt{a(b + 2c + d)}, \ \sqrt{b(c + 2d + a)}, \ \sqrt{c(d + 2a + b)}, \ \sqrt{d(a + 2b + c)}$$

and

$$\sqrt{\frac{a}{b + 2c + d}}, \ \sqrt{\frac{b}{c + 2d + a}}, \ \sqrt{\frac{c}{d + 2a + b}}, \ \sqrt{\frac{d}{a + 2b + c}}.$$

5.3.7. (a) We have

$$
\frac{a}{b+c} + \frac{b}{c+d} + \frac{c}{d+a} + \frac{d}{a+b} = \frac{a^2}{a(b+c)} + \frac{b^2}{b(c+d)} + \frac{c^2}{(d+a)} + \frac{d^2}{d(a+b)}
$$

$$
\overset{(*)}{\geq} \frac{(a+b+c+d)^2}{ab+ac+bc+bd+cd+ac+ad+bd}
$$

$$
= \frac{a^2+b^2+c^2+d^2+2ab+2ac+2ad+2bc+2bd+2cd}{ab+bc+cd+da+2ac+2bd}
$$

$$
\overset{(**)}{\geq} \frac{2ac+2bd+2ab+2ac+2ad+2bc+2bd+2cd}{ab+bc+cd+da+2ac+2bd} = 2.
$$

Inequality (*) follows from CBS and (**) follows from Cauchy's inequalities (5.1.5).

(b) We have

$$
\frac{a+c}{a+b} + \frac{b+d}{b+c} + \frac{c+a}{c+d} + \frac{d+b}{d+a}
$$

$$
= \frac{(a+c)^2}{(a+c)(a+b)} + \frac{(b+d)^2}{(b+d)(b+c)} + \frac{(c+a)^2}{(c+a)(c+d)} + \frac{(d+b)^2}{(d+b)(d+a)}
$$

$$
\overset{(*)}{\geq} \frac{4(a+b+c+d)^2}{(a+c)(a+b) + (b+d)(b+c) + (c+a)(c+d) + (d+b)(d+a)} = 4,
$$

with (*) following from CBS.

5.3.8. We have

$$
\frac{a^3}{b+c+d} + \frac{b^3}{c+d+a} + \frac{c^3}{d+a+b} + \frac{d^3}{a+b+c}
$$

$$
= \frac{a^4}{a(b+c+d)} + \frac{b^4}{b(c+d+a)} + \frac{c^4}{c(d+a+b)} + \frac{d^4}{d(a+b+c)}
$$

$$
\overset{(*)}{\geq} \frac{(a^2+b^2+c^2+d^2)^2}{2(ab+ac+ad+bc+bd+cd)} = \frac{S^2}{2+k} \overset{(**)}{\geq} \frac{S}{2+k} \overset{(***)}{\geq} \frac{1}{3}.
$$

Here $S := a^2 + b^2 + c^2 + d^2$ and $k := 2(ac + bd)$;

- Inequality (∗) follows from CBS;
- Inequality (∗∗) follows from $S \geq ab + bc + cd + da = 1$;
- Inequality (∗ ∗ ∗) is true because $S \geq 1$ and $S \geq 2ac + 2bd = k$, so $3S \geq 2 + k$. Alternatively, since $S \geq 1$ and $S \geq k$, if $k \leq 1$, then $\frac{S}{2+k} \geq \frac{1}{2+1} = \frac{1}{3}$, and if $k \geq 1$, then $\frac{S}{2+k} \geq \frac{k}{2+k} = \frac{1}{2+\frac{1}{k}} \geq \frac{1}{3}$.

5.3.11. (a) Let $x = 1/a$, $y = 1/b$, $z = 1/c$, $S := x + y + z$. Then

$$
\frac{1}{a^3(b+c)} = \frac{x^3yz}{y+z} = \frac{x^2}{S-x} = -x - S + \frac{S^2}{S-x}.
$$

(b) *Answer*: Either $\alpha \geq 1$ or $\alpha \leq -2$.

4. Geometric interpretation (3*)

5.4.1. (a) If
$$a + A = b + B = c + C = k, \quad \text{then} \quad aB + bC + cA \leq k^2.$$

(b) If $x_1, x_2, x_3, x_4 \leq 1$, then
$$x_1(1 - x_2) + x_2(1 - x_3) + x_3(1 - x_4) + x_4(1 - x_1) \leq 2.$$

5.4.2. (a) At what value of x does the expression $\sqrt{x^2 + 1} + \sqrt{(x - 1)^2 + 4}$ attain its smallest value?

(b) Find the smallest value of
$$\sqrt{x^2 + 1} + \sqrt{y^2 + 4} + \sqrt{z^2 + 9} + \sqrt{t^2 + 16}$$
subject to the condition $x + y + z + t = 17$.

(c) Prove that if $a, b, c > 0$, then
$$\sqrt{a^2 - ab + b^2} + \sqrt{b^2 - bc + c^2} \geq \sqrt{a^2 + ac + c^2}.$$

(d) If $\gamma = \sqrt{a^2 + b^2}$, $\beta = \sqrt{a^2 + c^2}$, and $\alpha = \sqrt{b^2 + c^2}$, then
$$\sqrt{(\alpha + \beta + \gamma)(\alpha + \beta - \gamma)(\alpha - \beta + \gamma)(\beta + \gamma - \alpha)} = 2\sqrt{a^2 b^2 + b^2 c^2 + a^2 c^2}.$$

(e) The following inequality holds:
$$\sqrt{4a^2 + b^2 + c^2 + 4ab + 4ac - 2bc} + \sqrt{4b^2 + a^2 + c^2 + 4ab + 4bc - 2ac}$$
$$\geq \sqrt{4c^2 + a^2 + b^2 + 4ac + 4bc - 2ab}.$$

(f) The following inequality holds:
$$\sqrt{ab(a + b)} + \sqrt{bc(b + c)} + \sqrt{ca(c + a)} \geq \sqrt{(a + b)(b + c)(c + a)}.$$

5.4.3. Let
$$x, y, z > 0 \quad \text{and} \quad \begin{cases} x^2 & + & xy & + & \frac{y^2}{3} & = & 25, \\ & & & & \frac{y^2}{3} + z^2 & = & 9, \\ z^2 & + & zx & + & x^2 & = & 16. \end{cases}$$

Find $xy + 2yz + 3zx$.

5.4.4. (a) Let $a_0 = 1/3$ and $a_n = \sqrt{\frac{1 + a_{n-1}}{2}}$. Prove that the sequence $\{a_n\}$ is monotone.

(b) Prove that from any four numbers, you can choose two numbers x and y such that $0 \leq \frac{x - y}{1 + xy} \leq 1$.

(c) Find all $x > 0$ that satisfy
$$x(8\sqrt{1 - x} + \sqrt{1 + x}) \leq 11\sqrt{1 + x} - 16\sqrt{1 - x}.$$

(d) Solve the following system of equations:

$$
\begin{cases}
\cot x \cot y \; - \; 5 \; = \; \dfrac{\cos z}{\sin x \sin y}, \\[2mm]
\cot y \cot z \; + \; 11 \; = \; \dfrac{\cos x}{\sin y \sin z}, \\[2mm]
\cot z \cot x \; + \; 7 \; = \; \dfrac{\cos y}{\sin z \sin x}.
\end{cases}
$$

5.4.5. Find an explicit formula for x_n if $x_{n+1} = x_n \sqrt{\frac{x_n + x_{n-1}}{2x_{n-1}}}$ and
(a) $x_0 = 1$, $x_1 = 1/2$;
(b) $x_0 = 1$, $x_1 = 2$.

5.4.6. If $ab = 4$ and $c^2 + 4d^2 = 4$, then
(a) $(a - c)^2 + (b - d)^2 > 1.6$;
(b)* $\sqrt{(a - c)^2 + (b - d)^2} \geq \frac{4\sqrt{\alpha} - \sqrt{4+\alpha^2}}{\sqrt{1+\alpha^2}}$ for any $\alpha > 0$.

Suggestions, solutions, and answers

5.4.1. (a) Consider an equilateral triangle PQR with side k. Mark points K, L, and M on sides PQ, QR, and RP, respectively, so that $PK = A$, $QL = B$, and $RM = C$. Then it is easy to check that $KQ = a$, $LR = b$, and $MP = c$. Therefore

$$
(aB + bC + cA)\sin 60° = S_{KQL} + S_{LRM} + S_{MPK} < S_{PQR} = k^2 \sin 60°.
$$

The inequality holds since the triangles KQL, LRM, and MPK are contained in the triangle PQR and do not intersect each other. Canceling by $\sin 60°$, we get the required inequality.

5.4.2. (c) Consider segments of length a and c forming an angle $60°$ with a segment of length b.

(d) Consider three pairwise perpendicular segments of lengths a, b, and c which meet at the same point in space.

(e) First prove that one can build a triangle from the medians of a triangle. Apply this result to a triangle with sides $a + b$, $b + c$, and $c + a$.

5.4.4. (a) Use the identity $\cos(\alpha/2) = \pm\sqrt{\frac{1+\cos\alpha}{2}}$.

(c) Divide by $1 - x$ and make the substitution $u = \sqrt{\frac{1+x}{1-x}}$ or (really the same substitution) $x = \cos 2t$.

(d) This problem requires hyperbolic geometry on the pseudosphere, for those familiar with the subject.

Chapter 6

Sequences and limits

This chapter is almost independent of other parts of the book. In other chapters, we will only use simple facts from this chapter.

1. Finite sums and differences (3)

The sequence $b_n = \Sigma a_n := a_1 + \cdots + a_n$ is said to be a *sequence of sums* of the sequence $\{a_n\}_{n=1}^{\infty}$, and the sequence $c_n = \Delta a_n := a_{n+1} - a_n$ is said to be its *sequence of differences*.

For example, $\Delta 2^n = 2^n$ and $\Sigma 2^n = 2^{n+1} - 2$. (The sum and difference are analogues of the integral and derivative.) In this section, the variable n denotes the index of the sequence element for which the sum or difference is taken. Thus $\Delta 2^k = 0$, since 2^k is a constant function of the variable n.

6.1.1. Find
    (a) Δn^k for every integer $k \geq -1$;   (b) $\Delta \cos n$;   (c) $\Delta(n \cdot 2^n)$.

6.1.2. Find
    (a) $\Sigma \sin n$;   (b) $\Sigma \frac{1}{n(n+1)\cdots(n+k)}$ for a positive integer k.

6.1.3. Which of the following equalities hold for some non-constant sequence a_n?
    (a) $\Delta a_n = 0$;   (b) $\Delta a_n = 1$;   (c) $\Delta a_n = a_n$;
    (d) $\Sigma a_n = a_n$;   (e) $\Sigma \Delta a_n = a_n$;   (f) $\Delta \Sigma a_n = a_n$.

Define the kth difference Δ^k of a sequence $\{a_n\}$ to be $\Delta(\Delta(\cdots))a_n$, where the difference operation is applied k times.

6.1.4. (a) Find $\sum_{k=0}^{n} (-1)^k k^2 \binom{n}{k}$.

(b) **Lemma.** The kth difference of a polynomial of degree k is a constant, and the $(k+1)$th difference is 0.

(c) (Challenge.) Express $\Delta^k a_n$ in terms of $a_n, a_{n+1}, \ldots, a_{n+k}$.

(d) **Lemma.** The equality $\Delta^k a_n = 0$ holds if and only if a_n is a polynomial in n of degree not greater than $k - 1$.

(e) There exists a polynomial $P_\lambda(n)$ of degree l for $\lambda \neq 1$ and degree $l-1$ for $\lambda = 1$ such that $\Delta(n^l\lambda^n) = P_\lambda(n)\lambda^n$.

(f) **Leibniz formula.** The following equality holds:

$$\Delta(a_n b_n) = a_{n+1}\Delta b_n + b_n\Delta a_n.$$

(g)* Formulate and prove a similar formula for $\Delta^l(a_n b_n)$.

6.1.5. (a) Find Σn^k for $k = 0, 1, 2, 3, 4$.

(b) **Lemma.** The sequence of sums of a polynomial of degree $k \geq 0$ is a polynomial of degree $k+1$.

6.1.6. (a) Find $\Sigma(n \cdot 2^n)$.

(b) Let $\mathbb{R}[x]$ denote the set of polynomials in the variable x with coefficients in $\mathbb{R}$. For any polynomial $f \in \mathbb{R}[x]$ and any number $\lambda \in \mathbb{R}$, there exist a polynomial $g \in \mathbb{R}[x]$ and a number $C \in \mathbb{R}$ such that for any $n \geq 1$ we have

$$\Sigma(f(n)\lambda^n) = g(n)\lambda^n + C \quad \text{and} \quad \deg g(n) = \begin{cases} \deg f(n), & \lambda \neq 1, \\ \deg f(n) + 1, & \lambda = 1. \end{cases}$$

(c) If $\Delta^l b_n = \lambda^n n^k$ for non-negative l, k, and λ, then $b_n = g(n)\lambda^n + h(n)$ for some polynomials h and g, where h has degree less than l and g has degree not greater than $l+k$ if $\lambda = 1$ and not greater than k if $\lambda \neq 1$.

6.1.7. The Abel summation formula (an analogue of integration by parts for sums). Formulate and prove the formula for the sum of products, which is obtained by summing the Leibniz formula 6.1.4 (f).

Hints

6.1.2. (b) Start with $k = 1, 2$; decompose the fraction into simplest fractions.

6.1.4. (a) Use (b) and (c).

(b) The statement follows from the solution of problem 6.1.1 (a).

(c) Verify that $\Delta^k a_n = \sum_{j=0}^{k}(-1)^j\binom{k}{j}a_{n+j}$.

(d) In the "if" direction, the result follows from part (b), and thus from the solution to problem 6.1.1 (a). In the "only if" direction the result follows from the solution to problem 6.1.1 (a); compare with problem 6.1.5 (a).

(e) Apply induction on l using the solution of 6.1.1 (a) and part (f) of this problem.

(g) Verify that $\Delta^l(a_n b_n) = \sum_{j=0}^{l}(-1)^j\binom{l}{j}\Delta^j a_{n+j}\Delta^{l-j}b_n$.

6.1.5. (a) We have $\Delta n^{k+1} = (k+1)n^k + \ldots$, which implies $n^{k+1} = (k+1)\Sigma n^k + \Sigma(\ldots)$.

(b) Apply induction on k using the solution of problem 6.1.1 (a).

6.1.6. (a) This is similar to problem 6.1.5 (a). We have $\Delta(n^2 \cdot 2^n) = n \cdot 2^n + \ldots$, so $n^2 \cdot 2^n = \Sigma(n \cdot 2^n) + \Sigma(\ldots)$.

(b) This is similar to (a) and problem 6.1.5.

6.1.7. See problem 6.5.6 (b).

Suggestions, solutions, and answers

6.1.1. (a) For $k > 0$ we have

$$\Delta n^k = (n+1)^k - n^k = \binom{k}{1}n^{k-1} + \binom{k}{2}n^{k-2} + \ldots + \binom{k}{k-1}n + 1.$$

For $k = 0$ we have $\Delta n^0 = 0$. For $k = -1$ we have $\Delta\frac{1}{n} = -\frac{1}{n(n+1)}$.

(b) We have $\Delta \cos n = \cos(n+1) - \cos n = -2\sin\frac{1}{2}\sin\left(n+\frac{1}{2}\right)$.

(c) We have $\Delta(n2^n) = (n+1)2^{n+1} - n2^n = (n+2)2^n$.

6.1.2. (a) *Answer:* $\dfrac{\cos\frac{1}{2} - \cos\left(n+\frac{1}{2}\right)}{2\sin\frac{1}{2}}$.

Solution. According to 6.1.1 (b) we have

$$\Delta \cos n = -2\sin\frac{1}{2}\sin\left(n+\frac{1}{2}\right), \quad \text{so} \quad \Delta\cos\left(n-\frac{1}{2}\right) = -2\sin\frac{1}{2}\sin n.$$

Summing (applying Σ to) both sides of the equality, we get

$$\cos\left(n+\frac{1}{2}\right) - \cos\frac{1}{2} = \Sigma\left(-2\sin\frac{1}{2}\sin n\right).$$

(b) *Answer:* $\frac{1}{(k+2)(k+2)!} + \frac{1}{(k+1)!} - \frac{1}{(k+2)n(n+1)\cdots(n+k+1)}$.

Solution. Verify that the following equality holds for positive integers k:

$$\Delta\frac{1}{n(n+1)\cdots(n+(k+1))} = -\frac{k+2}{(n+1)(n+2)\cdots(n+k+1)}.$$

Summing (applying Σ to) both sides of equality yields

$$\frac{1}{(n+1)(n+2)\cdots(n+k+2)} - \frac{1}{(k+2)!}$$

$$= -(k+2)\Sigma\frac{1}{(n+1)(n+2)\cdots(n+k+1)}.$$

Note that the mth term of the sequence

$$\Sigma\frac{1}{(n+1)(n+2)\cdots(n+k+1)}$$

is equal to the $(m+1)$th term of the sequence

$$-\frac{1}{(k+1)!} + \Sigma\frac{1}{n(n+1)\cdots(n+k)}.$$

Therefore

$$\frac{1}{n(n+1)\cdots(n+k+1)} - \frac{1}{(k+2)!}$$

$$= -(k+2)\left(-\frac{1}{(k+1)!} + \Sigma\frac{1}{n(n+1)\cdots(n+k)}\right).$$

Finally, divide by $k+2$.

2. Linear recurrences (3)

Thanks to T. Takebe for helpful comments.

In the following problems, the word "find" means "find as a formula containing polynomials in n, a^n, and $\cos(\omega n + \varphi)$."

6.2.1. Find the number of tilings with dominoes, that is, 1×2 rectangles, of

(a) a $2 \times n$ rectangle;   (b) a $3 \times 2n$ rectangle.

Here tilings that differ by rotation or reflection are considered distinct.

See also [**RSG+16**, problems 1.1.3 and 6.1.1 (d, e)].

6.2.2. Which of the following sequences satisfies the recurrence relation $a_{n+2} - 2a_{n+1} + a_n = 0$?

(a) $a_n = 5n + 3$;   (b) $a_n = (2n+1) \cdot 2^n$;   (c) $a_n = \cos(2n)$.

6.2.3. Find all sequences $\{a_n\}$ with $a_1 = 1$ and $a_2 = 3$ satisfying the following recurrence relations for all $n \geq 1$:

(a) $a_{n+2} = 3a_{n+1} - 2a_n$; (b) $a_{n+2} = 5a_{n+1} - 6a_n$;
(c) $a_{n+2} = 2a_{n+1} - a_n$; (d) $a_{n+2} = 4a_{n+1} - 4a_n$;
(e) $a_{n+2} = a_{n+1} - a_n$;   (f)* $a_{n+3} = 6a_{n+2} - 11a_{n+1} + 6a_n$.

6.2.4. Find all sequences such that $a_1 = 5$ and $a_{n+1} - 2a_n$ is equal to

(a) 0;   (b) 1;   (c) n;   (d) 3^n;   (e)* 2^n;   (f)* $n \cdot 3^n$.

6.2.5. The same question as above, replacing $a_{n+1} - 2a_n$ with $a_{n+2} - 5a_{n+1} + 6a_n$.

6.2.6. (a) **Theorem.** If λ is a root of multiplicity l of the equation $x^k = p_{k-1}x^{k-1} + \cdots + p_0$, then for any polynomial $f \in \mathbb{R}[x]$ of degree less than

l, the function $f(n)\lambda^n$ is a solution to the linear homogeneous recurrence relation $a_{n+k} = p_{k-1}a_{n+k-1} + \cdots + p_0 a_n$.

(b) For any $l_1, \ldots, l_k \in \mathbb{Z}$ and distinct $\lambda_1, \ldots, \lambda_k \in \mathbb{C}$, the sequences $n^i \lambda_j^n$, $j = 1, \ldots, k$, $i = 0, \ldots, l_j$, are linearly independent.

(c)* **Theorem.** Let $p_{k-1}, \ldots, p_0$ satisfy $a_{n+k} = p_{k-1}a_{n+k-1} + \cdots + p_0 a_n$ for each n, and let $\lambda_1, \ldots, \lambda_k$ be distinct complex roots of the polynomial $x^k - p_{k-1}x^{k-1} - \cdots - p_1 x - p_0$ with multiplicities $l_1, \ldots, l_k$. Then there exist polynomials $f_1, \ldots, f_k$ such that $\deg f_j < l_j$ and $a_n = f_1(n)\lambda_1^n + \cdots + f_k(n)\lambda_k^n$ for any n.

(d)* (Challenge.) Formulate and prove a theorem about the explicit form of the solutions of kth-order linear non-homogeneous recurrence relations.

For a more advanced interpretation and application of the method of variation of parameters, see [**VSY17**].

The concept of derivative used below is defined for polynomials in section 2 of Chapter 7 and for the general case in, for example, [**Zor15**].

6.2.7. Find all differentiable functions $y\colon \mathbb{R} \to \mathbb{R}$ satisfying $y(0) = 1$ and $y'(x) - 2y(x) = f(x)$ for all x, where
  (a) $f(x) = 0$;   (b) $f(x) = 1$;   (c) $f(x) = x$;   (d) $f(x) = e^x$;
  (e)* $f(x) = e^{2x}$;   (f)* $f(x) = xe^x$.

6.2.8. The same problem as above, replacing $y'(x) - 2y(x)$ with $y''(x) - 5y'(x) + 6y(x)$.

6.2.9. (a) **Theorem.** If λ is a root of multiplicity l of $x^k = p_{k-1}x^{k-1} + \cdots + p_0$, then for any polynomial $f \in \mathbb{R}[x]$ of degree less than l the function $f(x)e^{\lambda x}$ is a solution of the linear homogeneous differential equation $y^{(k)} = p_{k-1}y^{(k-1)} + \cdots + p_0 y$.

(b)* Formulate and prove a theorem about the explicit form of solutions of a linear homogeneous differential equation $y^{(k)} = p_{k-1}y^{(k-1)} + \cdots + p_0 y$ of order k.

(c)* Same as above, but for linear non-homogeneous differential equations.

Hints

6.2.1. (a) These are the Fibonacci numbers.

6.2.3. (a) For $c_n := a_{n+1} - a_n$, we get $c_{n+1} = 2c_n$.

(a–f) *The method of variation of parameters.* Find the solution (not taking into account the initial conditions) in the form $a_n = \lambda^n$ and consider $b_n = a_n/\lambda^n$ and $c_n = b_{n+1} - b_n$.

(b) The sequences 2^n and 3^n satisfy the recurrence relation (not taking into account the initial conditions). Let $b_n := a_n/2^n$ and $c_n = b_{n+1} - b_n$. Then $c_{n+1} = 3c_n$.

6.2.4. See problem 6.2.3.

6.2.6. See problem 6.2.3.
  (b) Consider the limit as $n \to \infty$.

6.2.7. See problem 6.2.3 with $y(x) = z(x)e^{\lambda x}$.

Suggestions, solutions, and answers

Answers provided by A. Khrabrov.

6.2.3. *Answers:*   (a) $2^n - 1$;   (b) 3^{n-1};   (c) $2n - 1$;
  (d) $(n+1)2^{n-2}$;   (e) $\frac{a-7}{6}3^n + \frac{9-a}{2}(2^n - 1)$, $a := a_3$.

6.2.4. *Answers:*   (a) $5 \cdot 2^{n-1}$;   (b) $3 \cdot 2^n - 1$;   (c) $7 \cdot 2^{n-1} - n - 1$;
  (d) $2^n + 3^n$;   (e) $(n+4)2^{n-1}$;   (f) $(n-3)3^n + 11 \cdot 2^{n-1}$.

6.2.5. *Answers,* where $a := a_2$:
  (a) $(a - 10)3^{n-1} + (15 - a)2^{n-1}$;
  (b) $\left(a - \frac{19}{2}\right)3^{n-1} + (14 - a)2^{n-1} + \frac{1}{2}$;
  (c) $\left(a - \frac{37}{4}\right)3^{n-1} + (13 - a)2^{n-1} + \frac{n}{2} + \frac{3}{4}$;
  (d) $(n + a - 14)3^{n-1} + (18 - a)2^{n-1}$;
  (e) $(a - 8)3^{n-1} + (14 - a - n)2^{n-1}$;
  (f) $\left(\frac{n^2 - 7n}{2} + a - 1\right)3^{n-1} + (9 - a)2^{n-1}$.

3. Concrete theory of limits (4*)

The problems of this section provide an interesting way to approach the theory of limits. Similar problems using these estimation methods often come up in olympiads and in applied and theoretical mathematics.

When solving these problems, you may not use functions such as $\sqrt[n]{x}$, a^x, $\log_a x$, $\arcsin x$, etc. without first defining them. However, in order to define them you may need to prove the existence of x such that $x^2 = 2$. In this case you would first need to solve the corresponding suggested problem! An exception is that if a particular function is used in the statement of the problem, then it can be used in the solution. You can also use the various properties of inequalities without proof.

6.3.1. Find at least one N such that for any $n > N$, the inequality $a_n > 10^9$ holds in the case where

(a) $a_n = \sqrt{n}$;

(b) $a_n = n^2 - 3n + 5$;

(c) $a_n = 1.02^n$;

(d) $a_n = 1 + \frac{1}{2} + \frac{1}{3} + \frac{1}{4} + \cdots + \frac{1}{n}$.

6.3.2. Bernoulli's inequality. Prove that $(1+x)^a \geq 1 + ax$ for any $x \geq -1$ and

(a) integer $a \geq 1$;

(b) rational $a \geq 1$;

(c) real $a \geq 1$.

6.3.3. Find at least one pair (a, N) such that for any $n > N$, the inequality $|a_n - a| < 10^{-8}$ holds for

(a) $a_n = \frac{n^2 - n + 28}{n - 2n^2}$;

(b) $\sqrt{5 + \dfrac{2}{n}}$;

(c) $a_n = n\left(\sqrt{1 + \frac{1}{n}} - 1\right)$;

(d) $a_n = n\left(\sqrt[3]{1 + \frac{1}{n}} - 1\right)$;

(e) $a_n = 0.99^n$;

(f) $a_n = \sqrt[n]{2}$;

(g) $a_n = n^9 / 2^n$;

(h)* $a_n = (1 + 1/n)^n$;

(i)* $a_n = n(\sqrt[n]{2} - 1)$;

(j) $a_n = \frac{1}{1^2} + \frac{1}{2^2} + \cdots + \frac{1}{n^2}$;

(k)* $a_n = \frac{1}{1\sqrt{1}} + \frac{1}{2\sqrt{2}} + \frac{1}{3\sqrt{3}} + \cdots + \frac{1}{n\sqrt{n}}$;

(l)* $a_n = \frac{1}{0!} + \frac{1}{1!} + \cdots + \frac{1}{n!}$;

(m)* $a_n = 1 - \frac{1}{3} + \frac{1}{5} - \frac{1}{7} + \cdots + \frac{(-1)^n}{2n+1}$.

6.3.4. Find at least one pair (a, δ) with $\delta > 0$ such that for any $x \in (-\delta, \delta)$, the inequality $|f(x) - a| < 3 \cdot 10^{-9}$ holds for $f(x)$ equal to

(a) x^3;

(b) 3^x;

(c) $\sin x$;

(d) $\frac{\sin x}{x}$;

(e) $\frac{\sqrt{1 + x^5}}{\cos x - 2}$;

(f) $\frac{1 + \sin x}{x^3 - 1}$;

(g) $(1 + 1/x)^x$.

Suggestions, solutions, and answers

6.3.1. (b) $n^2 - 3n + 5 > n(n-3) > n$ for any $n > 4$.
   (c) Use problem 6.3.2.

6.3.2. Use induction on k.

6.3.3. (a) $a = -\frac{1}{2}$;

   (b) $\sqrt{5 + \dfrac{2}{n}} - \sqrt{5} = \dfrac{\left(\sqrt{5 + \dfrac{2}{n}} - \sqrt{5}\right)\left(\sqrt{5 + \dfrac{2}{n}} + \sqrt{5}\right)}{\sqrt{5 + \dfrac{2}{n}} + \sqrt{5}}$

   $\qquad = \dfrac{2}{n\left(\sqrt{5 + \dfrac{2}{n}} + \sqrt{5}\right)}.$

   (e, f) Set $a = 0$ for (e) and $a = 1$ for (f), and use Bernoulli's inequality.
   (g) Put $a = 0$ and find N such that $(n+1)^9/n^9 < 1.5$ for any $n > N$.
   (h) Prove and use the following inequalities:

$$\left(1 + \frac{1}{n}\right)^n < \left(1 + \frac{1}{n+1}\right)^{n+1} < \left(1 + \frac{1}{n+1}\right)^{n+2} < \left(1 + \frac{1}{n}\right)^{n+1}.$$

And then

$$\left(1 + \frac{1}{n}\right)^{n+1} - \left(1 + \frac{1}{n}\right)^n = \frac{1}{n}\left(1 + \frac{1}{n}\right)^n < \frac{4}{n}.$$

   (i) Use the $\log_2$ function. A sketch of the proof of continuity of $f(x) = 2^x$, which is necessary for its definition, was given in parts (e) and (f). The continuity of the function $f(x) = x^n$ with integer n, which is necessary for the definition of $f(x) = 2^x$, was actually proved in parts (c) and (d).
   For $L := \log_2\left(1 + \frac{1}{n}\right)$ we have

$$n\left(\sqrt[n]{2} - 1\right) = n\left(\sqrt[n]{\left(1 + \frac{1}{n}\right)^{\frac{1}{L}}} - 1\right) = n\left(\left(1 + \frac{1}{n}\right)^{\frac{1}{nL}} - 1\right) \leq \frac{1}{nL}.$$

The inequality can be verified using arguments similar to those in the proof of Bernoulli's inequality for $a < 1$. Using $(1 + x)^a \geq 1 + ax + a(a-1)x^2$ for $a < 1$, we get a sharp estimation in the other direction. The value of nL was estimated in part (g).
   (i–l) The number a should not necessarily be equal to the *limit*.

6.3.4. (a) If $|x| < 1$ then $|x^3| < |x|$.
   (c) Use the inequality $\sin x < x$.
   (e, f) If

$$|f(x) - a| < \varepsilon/2 \quad \text{when} \quad x \in (-\delta_1, \delta_1) \quad \text{and}$$

$$|g(x) - b| < \varepsilon/2 \quad \text{when} \quad x \in (-\delta_2, \delta_2),$$

$$\text{then} \quad |f(x) + g(x) - a - b| < \varepsilon \quad \text{when} \quad x \in (-\min\{\delta_1, \delta_2\}, \min\{\delta_1, \delta_2\}).$$

The same is true when the sum is replaced by the difference. Similar statements are also true when the sum is replaced with a product or quotient.

4. How does a computer calculate the square root? (4*) By A. C. Vorontsov and A. I. Sgibnev

The goal of the problems below is to show how to calculate the square root to any precision using only arithmetic operations (for example, with a simple calculator). The most difficult and interesting problem is the error estimation (such estimations were actually carried out in the previous section).

We quote from Heron's text. He explains his method with an example: finding the square root of 720.

> *Since 720 does not have a rational root, we will find the square root with a very small error as follows. Since the nearest integer square is 729 and it has a root equal to 27, divide 720 by 27. You get $26\frac{2}{3}$. Add 27. You get $53\frac{2}{3}$. Take the half of that. You get $26\frac{5}{6}$. Thus, the nearest root of 720 will be $26\frac{5}{6}$. If you multiply it by itself you will get $720\frac{1}{36}$, so the error is equal to a 36th part of the unit. If we would like the error to become a smaller part of the unit than the 36th, then instead of 729 we take the newly found number $720\frac{1}{36}$ and, having done the same, we find that the error has become much less than $\frac{1}{36}$.* [**Vyg67**]

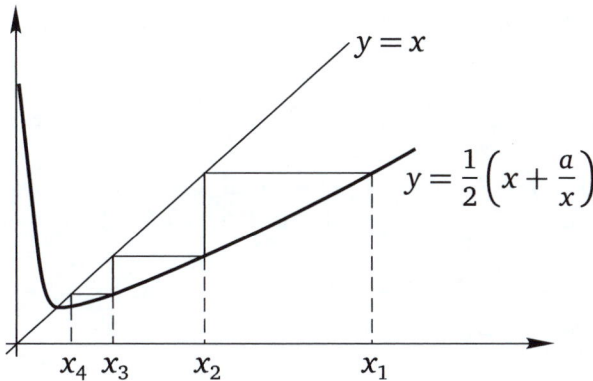

FIGURE 6.1. Heron's method

Let us write Heron's calculations in modern notation (Figure 6.1). For any $x_1 \neq 0$ and $a > 0$ define the sequence by the formula

$$x_{n+1} = \frac{1}{2}\left(x_n + \frac{a}{x_n}\right).$$

6.4.1. This definition makes sense, since $x_n \neq 0$ for any n.

6.4.2. This problem, and therefore the concept of limit, is not formally used further.

The number A is called *limit* of the sequence a_n if for any $\varepsilon > 0$ there exists N such that for any $n > N$ the inequality $|A - a_n| < \varepsilon$ holds. We use the notation $A = \lim\limits_{n \to \infty} a_n$.

The following principle, due to Weierstrass, can be used without proof: *Any monotone bounded sequence has a limit.* Compare with section 5.

Find $\lim\limits_{n \to \infty} x_n$. You can start with $a = 2$ and then consider the general case.

6.4.3. (a) Let $a = 2$ and $x_1 = 1$, Find at least one, not necessarily minimal, N such that if $n > N$ we have $|x_n - \sqrt{2}| < 10^{-5}$.

(b, c, d, e) Same problem, but with $x_1 = 10, 100, 1000, 10^k$.

(a,'b') Same problem, but with $x_1 = -1, -10$.

6.4.4. If $a > 0$ then $|x_{n+1} - \sqrt{a}| \leq \frac{1}{2}|x_n - \sqrt{a}|$ for any $n \geq 2$. In other words, at each step of Heron's method, starting from the second, the error is reduced by at least a factor of two compared to the error at the previous step.

6.4.5. (Challenge.) To calculate cube roots, one can devise analogues of Heron's method, for example,

$$ y_{n+1} = \frac{1}{2}\left(y_n + \frac{a}{y_n^2}\right), \quad z_{n+1} = \frac{1}{3}\left(2z_n + \frac{a}{z_n^2}\right). $$

Find the rates of convergence of these sequences (for example, for $a = 8$); that is, formulate and prove analogues of problem 6.4.4.

The formula for z_n can be obtained using Newton's method; see [**Sgi09**].

Suggestions, solutions, and answers

6.4.2. *Answer*: The sequence converges to $\sqrt{a}$ when $x_0 > 0$ and to $-\sqrt{a}$ when $x_0 < 0$.

Outline of the solution. Let $x_0 > 0$. We prove that the sequence x_n is decreasing and is bounded below, which implies that it has a limit.

Note that $t + \frac{a}{t}$ cannot be too small:

$$ t + \frac{a}{t} = t - 2\sqrt{a} + \frac{a}{t} + 2\sqrt{a} = \left(\sqrt{t} - \frac{\sqrt{a}}{\sqrt{t}}\right)^2 + 2\sqrt{a} \geq 2\sqrt{a}. $$

Thus, $x_n \geq \sqrt{a}$ for any $n > 0$.

Next, we estimate the difference of the neighboring terms of the sequence.

$$x_{n+1} - x_n = \frac{1}{2}\left(\frac{a}{x_n} - x_n\right) = \frac{1}{2}\left(\frac{a - x_n^2}{x_n}\right) \leq 0.$$

Thus, the sequence x_n has a limit. Denote it by m. To find its value we pass to the limit on the left- and right-hand sides of the equality $x_{n+1} = \frac{1}{2}\left(x_n + \frac{a}{x_n}\right)$. We obtain $m = \frac{1}{2}\left(m + \frac{a}{m}\right)$. Therefore, $m = \pm\sqrt{a}$.

6.4.4. We have

$$2|x_{n+1} - \sqrt{a}| = \left|x_n + \frac{a}{x_n} - 2\sqrt{a}\right| = \left|(x_n - \sqrt{a}) + \left(\frac{a}{x_n} - \sqrt{a}\right)\right|$$

$$= \left|(x_n - \sqrt{a}) + \frac{\sqrt{a}}{x_n}(\sqrt{a} - x_n)\right| = \left|\left(1 - \frac{\sqrt{a}}{x_n}\right)(x_n - \sqrt{a})\right| \leq 1 \cdot |x_n - \sqrt{a}|.$$

The last inequality holds because

$$x_n = \frac{1}{2}\left(x_{n-1} + \frac{a}{x_{n-1}}\right) \geq \sqrt{a} \quad \text{for any } n \geq 2$$

(compare with problem 6.4.2).

6.4.5. The sequence z_n has the same rate of convergence as the sequence x_n. This can be proven analogously to 6.4.4. The sequence y_n converges much slower.

5. Methods of series summation (4*)

Newton regarded the concepts of differentiation and integration not as his main achievement, but as merely a natural language for writing the differential equations that express the laws of nature. Newton believed that his fundamental contribution was the method of solving differential equations using power series. We turn to this topic now.

(The expression "for any n" is often omitted.)

Let $a_n \geq 0$. The number A is called the *sum* of the series associated with the sequence $\{a_n\}$ if

1) $A \geq a_1 + \cdots + a_n$ for any n, and

2) for any $\varepsilon > 0$ there exists n such that $A < a_1 + \cdots + a_n + \varepsilon$.

For most of the problems in this section, the definition given above is sufficient and one does not need the following more general definition, where it is no longer assumed that $a_n \geq 0$: A is called the *sum* of the series $\{a_n\}$ if for any $\varepsilon > 0$ there exists N such that for any $n > N$ the inequality $|a_1 + \cdots + a_n - A| < \varepsilon$ holds.

Notation: $A = \sum_{n=1}^{\infty} a_n$, or simply $A = \sum a_n$. If a series has a sum, then the series is *convergent*; otherwise it is *divergent*. The number $A_n := a_1 + \cdots + a_n$ is called the nth partial sum of the series $\{a_n\}$.

In this and the following sections we will not need a rigorous theory of real numbers. (See, for example, the book [**Zor15**].) You may use without proof only the algebraic properties of real numbers and the following principle due to Weierstrass: The series $\{a_n\}$ of positive terms converges if its partial sums are bounded, i.e., there exists a number A with property 1). (This principle can be understood by considering infinite decimal expansions.)

In the following problems, equalities of series are understood in the sense that if the right-hand side exists then the left-hand side also exists and is equal to the right-hand side.

6.5.1. If $a_n \geq 0$ and $b_n \geq 0$, then

(a) $\sum\limits_{n=1}^{\infty} (a_n + b_n) = \sum\limits_{n=1}^{\infty} a_n + \sum\limits_{n=1}^{\infty} b_n$;   (b) $\sum\limits_{n=1}^{\infty} \lambda a_n = \lambda \sum\limits_{n=1}^{\infty} a_n$;

(c) $\sum\limits_{n=1}^{\infty} a_n = a_1 + \cdots + a_k + \sum\limits_{n=1}^{\infty} a_{k+n}$.

6.5.2. Explicit calculation of partial sums. Find

(a) $\sum\limits_{n=1}^{\infty} \frac{1}{n(n+1)}$;   (b) $\sum\limits_{n=1}^{\infty} \frac{1}{n(n+2)}$;

(c) $\sum\limits_{n=1}^{\infty} \frac{1}{(3n-1)(3n+2)}$;   (d)* $\sum\limits_{n=1}^{\infty} \frac{1}{n(n+1)(n+2)}$;

(e)* $\sum\limits_{n=1}^{\infty} \frac{2n+1}{n(n+1)(n+2)}$; (f)* $\sum\limits_{n=1}^{\infty} \frac{1}{n(n+1)(n+2)...(n+k)}$.

The sum $S = \sum\limits_{n=0}^{\infty} \frac{1}{2^n}$ can be found using the equation $1 + \frac{S}{2} = S$ once we have proven that this sum exists.

6.5.3. Using equalities. Find (a) $\sum\limits_{n=1}^{\infty} \frac{n}{2^n}$; (b)* $\sum\limits_{n=1}^{\infty} \frac{n^2}{2^n}$.

The sum $\sum\limits_{n=1}^{\infty} \frac{n}{2^n}$ can be found after proving the equality

$$\frac{1}{2} + \frac{1}{4} + \frac{1}{4} + \frac{1}{8} + \frac{1}{8} + \frac{1}{8} + \ldots = \left(\frac{1}{2} + \frac{1}{4} + \frac{1}{8} + \ldots\right) + \left(\frac{1}{4} + \frac{1}{8} + \ldots\right) + \left(\frac{1}{8} + \ldots\right) + \ldots$$

6.5.4. Regrouping terms.

(a) Find the sum $1 + 2x + 3x^2 + 4x^3 + \ldots$ after determining for which x the series converges.

(b) If $a_n \geq 0$ and $\sigma\colon \{0, 1, 2, \ldots\} \to \{0, 1, 2, \ldots\}$ is a permutation, that is, a one-to-one and onto mapping, then $\sum a_{\sigma(n)} = \sum a_n$.

(c) Find $\sum\limits_{n=k}^{\infty} \frac{n}{2^n}$.

The sum $\sum\limits_{n=1}^{\infty} \frac{n}{2^{n-1}}$ can be found after proving the equality

$$\left(1+\frac{1}{2}+\frac{1}{4}+\ldots\right)\left(1+\frac{1}{2}+\frac{1}{4}+\ldots\right) = 1+\frac{1}{1\cdot 2}+\frac{1}{2\cdot 1}+\frac{1}{4\cdot 1}+\frac{1}{2\cdot 2}+\frac{1}{1\cdot 4}+\ldots$$

6.5.5. Multiplication of series.

(a) Find $\sum\limits_{n=1}^{\infty} \frac{n(n+1)}{2^n}$.

(b) Prove the equality

$$\left(\frac{1}{0!}-\frac{1}{1!}+\frac{1}{2!}-\ldots+\frac{(-1)^n}{n!}+\ldots\right)\left(\frac{1}{0!}+\frac{1}{1!}+\ldots+\frac{1}{n!}+\ldots\right) = 1.$$

Be cautious: there are negative terms.

(c) Prove the equality

$$\left(\frac{1}{0!}+\frac{1}{1!}+\frac{1}{2!}+\ldots+\frac{1}{n!}+\ldots\right)^2 = \left(\frac{2^0}{0!}+\frac{2^1}{1!}+\ldots+\frac{2^n}{n!}+\ldots\right).$$

(d) If $a_n \geq 0$ and $b_n \geq 0$, then

$$\left(\sum_{n=0}^{\infty} a_n\right)\left(\sum_{n=0}^{\infty} b_n\right) = \sum_{n=0}^{\infty}(a_0 b_n + a_1 b_{n-1} + \ldots + a_n b_0).$$

The sum $\sum\limits_{n=1}^{\infty} \frac{n}{2^n}$ can be found using the equality

$$1\left(1-\frac{1}{2}\right)+2\left(\frac{1}{2}-\frac{1}{4}\right)+3\left(\frac{1}{4}-\frac{1}{8}\right)+\ldots = 1+\frac{1}{2}(2-1)+\frac{1}{4}(3-2)+\frac{1}{8}(4-3)+\ldots$$

6.5.6. Abel's summation formula.

(a)* Find $\sum\limits_{n=1}^{\infty} \frac{\cos\left(n+\frac{1}{2}\right)}{2^n}$.

(b) Prove the equality

$$\sum_{n=1}^{m} b_n(a_{n+1}-a_n) = a_{m+1}b_{m+1} - a_1 b_1 - \sum_{n=1}^{m} a_{n+1}(b_{n+1}-b_n).$$

What happens when $m = 1$?

(c) If the sequence $\{b_n\}$ is monotonic non-increasing and the sequence $\{a_n\}$ decreases monotonically to zero, then

$$\sum_{n=1}^{\infty} b_n(a_{n+1}-a_n) = -a_1 b_1 - \sum_{n=1}^{\infty} a_{n+1}(b_{n+1}-b_n).$$

6.5.7. The sum of an absolutely convergent series, that is, a series for which

$\sum_{n=1}^{\infty} |a_n|$ converges, does not depend on a permutation of the terms of the series.

6.5.8. One can rearrange the order of the terms in the sum $1 - \frac{1}{2} + \frac{1}{3} - \frac{1}{4} + \ldots + (-1)^{n+1}\frac{1}{n} + \ldots$ so that the sum of the new series becomes equal to
(a) ∞;   (b) 7.

6.5.9. Check whether an analogue of the statement of problem 6.5.5 (d) holds
(a) without the condition that $a_n, b_n > 0$;
(b) with the replacement of the condition $a_n, b_n > 0$ by absolute convergence.

6.5.10. Recomposition.
(a) Express $z^3 + 3z^2 - 2z - 1$ as a polynomial in $y = z + 1$.
(b) Find numbers a_n such that for any z with $|z| < 1$, we have $\frac{1}{z^2+2z+2} = \sum_{n=1}^{\infty} a_n z^n$.

6.5.11.* If you are familiar with derivatives, find $\sum_{n=1}^{\infty} n^k x^n$, where $|x| < 1$ and k is an integer.

6.5.12.* The sum $1 - \frac{1}{2} + \frac{1}{3} - \frac{1}{4} + \ldots$ is equal to the area of the curvilinear trapezoid bounded by the x-axis, the vertical lines $x = 1$ and $x = 2$, and the hyperbola $y = 1/x$.
(For the definition of the area, see, for example, [**Sko21**, section "The Dirichlet principle and its applications in geometry"].)

Hints

6.5.4. (b) Since $\sum_{n=1}^{k} a_{\sigma(n)} \leq \sum_{n=1}^{\max\{a_{\sigma(1)},\ldots,a_{\sigma(k)}\}} a_n$ for any n, we have $\sum a_{\sigma(n)} \leq \sum a_n$. Likewise, $\sum a_{\sigma(n)} \geq \sum a_n$.

6.5.7. Use 6.5.4 (b) and the equality $a_{n,\pm} := (a_n \pm |a_n|)/2$.

6.5.10. (b) Expand $\frac{1}{z^2+2z+2} = \frac{1}{(z+1)^2+1}$ into a series in powers of $z + 1$ and then rewrite it as a series in powers of z.

Suggestions, solutions, and answers

6.5.2. *Answers*:   (a) 1;   (b) 3/4;   (c) 1/6;   (d) 1/4;   (e) 5/4;   (f) $1/(k \cdot k!)$.
Hint. Expand into simple fractions.

6.5.3. *Answers*: (a) 2; (b) 6.

6.5.4. *Answers*: (a) $1/(1-x)^2$ when $0 \leq x < 1$; diverges when $x \geq 1$.
(c) $(k+1)/2^{k-1}$.
(a) *Hint for those familiar with derivatives.* Prove and use the equality

$$\left(\sum_{n=1}^{\infty} x^n \right)' = \sum_{n=1}^{\infty} nx^{n-1}.$$

6.5.5. (a) *Answer*: 8.

6.5.6. (a) *Answer*: $\dfrac{\cos \frac{1}{2} - 2\cos \frac{3}{2}}{4\cos 1 - 5}$.

6.5.10. (b) The first few terms are

$$\frac{1}{2} - \frac{1}{2}z + \frac{1}{4}z^2 - \frac{1}{8}z^4 + \frac{1}{8}z^5 - \frac{1}{16}z^6$$

$$+ \frac{1}{32}z^8 - \frac{1}{32}z^9 + \frac{1}{64}z^{10} - \frac{1}{128}z^{12} + \frac{1}{128}z^{13} + \dots$$

6.5.11. Prove and use the equality

$$\left(\sum_{n=1}^{\infty} n^k x^n \right)' = \sum_{n=1}^{\infty} n^{k+1} x^{n-1}.$$

6. Examples of transcendental numbers

6.A. Introduction (1)

A number x is called *transcendental* if it is not a root of an equation $a_t x^t + a_{t-1} x^{t-1} + \cdots + a_1 x + a_0 = 0$ with integer coefficients $a_t, a_{t-1}, \ldots, a_0$ and $a_t \neq 0$.

The first explicit example of a transcendental number was given by Joseph Liouville in 1835 (see Theorem 6.6.4 (a) and [**CR96**, Ch. 2, section 6]). In 1929, Kurt Mahler proved the transcendence of *Mahler's number*; see Theorem 6.6.7. This transcendence follows neither from the general theorem of Liouville 6.6.4 (b) nor from the theorems of Thue, Siegel, and Roth ([**CR96**, Ch. 2, section 6], [**Fel83**]). A more general result was obtained in [**Mah29**] (compare with [**Gal80**, **Nis96**]). However, the proof in [**Mah29**], as well as that in [**Nis96**], is long and difficult.

In this section, we will present simple proofs of the transcendence of the Liouville and Mahler numbers. The first of them is based on the elementary version of Lagrange's Mean Value Theorem 7.2.7 (c). Although it is known to specialists, unfortunately more complicated proofs are usually presented in classes. The second proof is based on the binary representation of numbers. Apparently, it was not known until [**Skoc**] and [**AS03**, section 13.3, pp. 399–401]. These proofs can be understood by high school students.

Note that there is a simple set-theoretic proof of the *existence* of transcendental numbers [**CR96**, Ch. 2, section 6]. It does not give an explicit example of a transcendental number, although it gives an algorithm for constructing one in decimal notation.

A preliminary version of part of this section was presented by A. Kaibkhanov at the 2002 international conference of Intel ISEF (Louisville, USA) and by I. Nikokoshev and A. Skopenkov at the Summer Conference of the Tournament of Cities (Beloretsk, Russia). We thank V. Volkov, A. Galochkin, D. Leshko, A. Pakharev, A. Rukhovich, and L. Shabanov for useful discussions.

Before studying this section it is useful to solve the problems in Chapter 3, section 1.

6.B. Problems (3*)

6.6.1. The following numbers are irrational:

(a) $e := \sum_{n=0}^{\infty} \frac{1}{n!}$;   (b) $\lambda := \sum_{n=0}^{\infty} 2^{-n!}$;   (c) $\mu := \sum_{n=0}^{\infty} 2^{-2^n}$.

The infinite sums used here are defined in Chapter 6, section 5.

6.6.2. None of the numbers e, λ, and μ is a root of a quadratic equation with integer coefficients.

6.6.3. For any rational number p/q that is not a root of a polynomial f of degree t with integer coefficients, the inequality $|f(p/q)| \geq q^{-t}$ holds.

Theorem 6.6.4 (Liouville). (a) The number λ is transcendental.

(b) For any polynomial of degree t with rational coefficients and an irrational root α, there exists $C > 0$ such that for any integers p and q, the inequality $\left|\alpha - \frac{p}{q}\right| > Cq^{-t}$ holds.

6.6.5. (a) The number μ is not a root of a cubic equation with integer coefficients.

(b) The equality $\mu^q = \sum_{n=0}^{\infty} d_n(q)2^{-n}$ holds, where $d_n(q)$ is the number of ordered representations of n as the sum of q powers of 2 (not necessarily distinct powers):

$$d_n(q) = |\{(w_1, \ldots, w_q) \in \mathbb{Z}^q \mid n = 2^{w_1} + \cdots + 2^{w_q} \text{ and } w_1, \ldots, w_q > 0\}|.$$

For example, $d_3(2) = 2$, since $3 = 2^0 + 2^1 = 2^1 + 2^0$. Define $d_0(0) := 1$.

Lemma 6.6.6. The number $d_n(q)$ of ordered representations of the number n as the sum of q powers of 2 does not exceed $(q!)^2$.

Theorem 6.6.7 (Mahler). The number μ is transcendental.

6.C. Proof of Liouville's Theorem (2)

First we prove that the number e is irrational (that is, we solve problem 6.6.1 (a)). Suppose, to the contrary, that there exists a linear polynomial $f(x) = bx + c$ with integer coefficients b and c, with $b \neq 0$, for which $f(e) = 0$. Let $e_s = \sum_{n=0}^{s} \frac{1}{n!}$. Since the equation $f(x) = 0$ has only one root, we have $f(e_s) \neq 0$. We get a contradiction from the following inequalities for $s = 2|b|$:

$$\frac{1}{s!} \leq |f(e_s)| = |f(e) - f(e_s)| = |b| \cdot (e - e_s) < \frac{2|b|}{(s+1)!}.$$

Next, write $\lambda_s = \sum_{n=0}^{s} 2^{-n!}$.

We will prove that the *Liouville number* λ *is irrational* (that is, we solve problem 6.6.1 (c).) Suppose, to the contrary, that there exists a linear polynomial $f(x) = bx + c$ with integer coefficients b and c, with $b \neq 0$, for which $f(\lambda) = 0$. Since the equation $f(x) = 0$ has only one root, $f(\lambda_s) \neq 0$. We get a contradiction from the following inequalities for $s = |b|$:

$$2^{-s!} \leq |f(\lambda_s)| = |f(\lambda) - f(\lambda_s)| = |b| \cdot (\lambda - \lambda_s) < 2|b| \cdot 2^{-(s+1)!}.$$

The first inequality holds since $f(\lambda_s) \neq 0$ can be represented as a fraction with denominator $2^{s!}$. The latter inequality follows from

$$\lambda - \lambda_s < 2^{-(s+1)!} \sum_{n=0}^{\infty} 2^{-n} = 2 \cdot 2^{-(s+1)!}.$$

Next, we show that λ is not a root of a quadratic polynomial $f(x) = ax^2 + bx + c$ with integer coefficients (that is, we solve problem 6.6.2 for λ; compare with subsection 3.A). Suppose, to the contrary, that λ is a root of such an equation. Since a quadratic equation has no more than two roots, we have $f(\lambda_s) \neq 0$ for sufficiently large s. Now for sufficiently large s we get a contradiction from the following inequalities:

$$2^{-2s!} \leq |f(\lambda_s)| = |f(\lambda) - f(\lambda_s)| = (\lambda - \lambda_s) \cdot |a(\lambda + \lambda_s) + b|$$
$$< (2|a|\lambda + |b|) \cdot 2 \cdot 2^{-(s+1)!}.$$

The first inequality holds since $f(\lambda_s) \neq 0$ can be represented as a fraction with denominator $2^{2s!}$. The second inequality is proved similarly to the linear case above.

Similar arguments work for e but *do not work* for μ.

Proof of Liouville's Theorem 6.6.4 (a). Suppose, to the contrary, that λ is a root of an algebraic equation $f(x) = a_t x^t + a_{t-1} x^{t-1} + \cdots + a_1 x + a_0 = 0$ with integer coefficients $a_0, \ldots, a_{t-1}, a_t$ with $a_t \neq 0$. Since such equation has only a finite number of roots, $f(\lambda_s) \neq 0$ for sufficiently large s. Then for sufficiently large s we get a contradiction with the following inequalities, whose verification is similar to the quadratic and linear cases above:

$$2^{-ts!} \leq |f(\lambda_s)| = |f(\lambda) - f(\lambda_s)|$$

$$= (\lambda - \lambda_s) \cdot \left| \sum_{0 \leq i < n \leq t} a_n \lambda^{n-1-i} \lambda_s^i \right| < C \cdot 2^{-(s+1)!}.$$

The first inequality holds because $f(\lambda_s) \neq 0$ can be represented as a fraction with denominator $2^{ts!}$. The second inequality is proved similarly to the case of a linear polynomial.   □

6.D. Simple proof of Mahler's Theorem (3*)

Let us demonstrate the idea of the proof using the following example. We prove that the base-10 number

$$\nu = \sum_{n=0}^{\infty} 10^{-2^n} = 0.11010001000000010\ldots_{10}$$

is not a root of a quadratic equation with integer coefficients. (Problem 6.6.2 for μ can be solved in the same way; consider the binary expansions of μ and μ^2.) Consider the decimal expansion of the number $-b\nu - c$ for integers b and c of the same sign (the case of different signs can be proved in a similar way). Consider nonzero digits in this decimal expansion located far enough from the decimal point. It is clear that they form "clusters" around the positions numbered 2^n, and each "cluster" represents the number b. For example, for $b = -17$ we have the following:

$$17\nu - c = \ldots.87170017000000170\ldots017\ldots.$$

However, in the base-10 expansion of

$$\nu^2 = \sum_{k,l=0}^{\infty} 10^{-2^k - 2^l} = 0.0121220\ldots122020002000000012\ldots_{10}$$

some nonzero digits are located near the $(2^k + 2^l)$th position, where $k \neq l$. But for sufficiently large k and l, the number $-b\nu - c$ will have zeros in these positions. Therefore $\nu^2 \neq -b\nu - c$.

Proof of Mahler's Theorem 6.6.7. Assume the converse: $f(\mu) := a_t \mu^t + a_{t-1}\mu^{t-1} + \cdots + a_1\mu + a_0 = 0$ for some integers $a_t, a_{t-1}, \ldots, a_0$ with $a_t \neq 0$. Expanding the brackets, we get

$$\mu^q = \left(\sum_{n=0}^{\infty} 2^{-2^n} \right)^q = \sum_{n=0}^{\infty} d_n(q) 2^{-n},$$

where $d_n(q)$ is the number of ordered representations of the number n as the sum of q powers of 2 (not necessarily distinct):

$$d_n(q) = |\{(w_1, \ldots, w_q) \in \mathbb{Z}^q \mid n = 2^{w_1} + \cdots + 2^{w_q} \text{ and } w_1, \ldots, w_q > 0\}|.$$

We have

$$f(\mu) = \sum_{n=0}^{\infty} d_n 2^{-n}, \quad \text{where} \quad d_n := a_t d_n(t) + a_{t-1} d_n(t-1) + \cdots + a_0 d_n(0).$$

It is clear that $d_n(q) = 0$ if and only if n has more than q ones in its binary expansion. For each p, let

- $k = k(p) := 2^{t+p}$;
- $m = m(p) := 2^p(2^t - 1)$, the greatest number less than k such that $d_m(t) \neq 0$;
- $s = s(p) := 2^p(2^t - 1) - 2^{p-1}$, the greatest number less than m such that $d_s(t) \neq 0$.

Then

$$\{2^s f(\mu)\} = \left\{ \sum_{n=0}^{\infty} d_n 2^{s-n} \right\} = \left\{ d_m 2^{s-m} + \sum_{n=k}^{\infty} d_n 2^{s-n} \right\}.$$

This expression is not equal to zero because

$$\left| \sum_{n=k}^{\infty} d_n 2^{s-n} \right| \overset{(1)}{<} |d_m| 2^{s-m} \overset{(2)}{<} 1/2.$$

By Lemma 6.6.6, there exists $D = D(f)$ such that $|d_n| \leq D$ for each n. Thus inequality (2) holds because $|d_m| 2^{s-m} \leq D \cdot 2^{-2^{p-1}} < 1/2$ for sufficiently large p. Inequality (1) holds because for sufficiently large p we have

$$\left| \sum_{n=k}^{\infty} d_n 2^{s-n} \right| \leq D \sum_{n=k}^{\infty} 2^{s-n} = D \cdot 2^{s+1-k} = D \cdot 2^{s-m+1-2^p}$$
$$< 2^{s-m} \leq |d_m| 2^{s-m}.$$

The latter inequality follows from $d_m(t) \neq 0$ and $d_m(q) = 0$ when $q < t$, so $d_m = a_t d_m(t) \neq 0$. $\qquad\square$

Proof of Lemma 6.6.6. (This proof was proposed by V. Volkov.)

We proceed by induction on q. For $q = 0$ we have $d_0(0) = 1 \leq 0!^2$. The inductive step follows from the inequality

$$d_n(q+1) \leq 1 + q^2 d_n(q).$$

Let us prove this inequality. Consider sequences $\vec{w} := (w_1, \ldots, w_{q+1})$ such that $n = 2^{w_1} + 2^{w_2} + \cdots + 2^{w_{q+1}}$. There is no more than one sequence $\vec{w}$ with distinct terms. In each $\vec{w}$ where not all terms are distinct, we replace two equal powers of 2 with their sum (a higher power of 2). Thus we get a new sequence $f(\vec{w}) = \vec{v} := (v_1, \ldots, v_q)$, for which $n = 2^{v_1} + 2^{v_2} + \cdots + 2^{v_q}$. It is possible that the function f is not one-to-one. A sequence $\vec{w}$ is obtained from $f(\vec{w})$ by splitting one of the powers of 2 into two and inserting the obtained new power of 2 some place to the right of the original one. The power of 2 for splitting can be selected in q ways. One can insert a new power of 2 some place to the right of the original one in less than q ways. Therefore, each sequence $\vec{v}$ for which $n = 2^{v_1} + 2^{v_2} + \cdots + 2^{v_q}$ has no more than q^2 preimages under f. This proves the necessary inequality.   □

The next problem is a good topic for research; see p. xviii. Parts (a), (b), and (c) are similar to Mahler's Theorem (6.6.7). The author does not know the solutions of (d) and (e), but certainly they are within the reach of a strong high school student (and may be known to specialists). Compare with [KaS06, Generalization].

6.6.8. Determine whether the number $\sum\limits_{n=0}^{\infty} a_n$ is transcendental, for

(a) $a_n = 2^{-3^n}$;

(b) $a_n = d_n 2^{-2^n}$ for some bounded sequence $d_n > 0$ of integers;

(c) $a_n = 2^{-f_n}$, where $f_{n+2} = f_{n+1} + f_n$ with $f_0 = f_1 = 1$ is the Fibonacci sequence;

(d)* $a_n = 2^{-[1.1^n]}$;

(e)* $a_n = n 2^{-2^n}$;

(f)* $a_n = 2^{n-2^n}$;

(g)* $a_n = (-1)^n 2^{-2^n}$.

Chapter 7

Functions

This chapter is almost independent of the rest of the book. Only simple facts from it are used elsewhere.

In this chapter, unless otherwise specified, a *polynomial* is a polynomial with *real* coefficients, and letters denote *real* numbers.

1. The graph and number of roots of a cubic polynomial

The author thanks M. Gorelov, A. Doledenok, M. Skopenkov, A. Sgibnev, and an anonymous reviewer of *Kvant* magazine for useful discussions.

1.A. Introduction

This subsection provides an elementary proof of the criterion for the existence of three distinct real roots of a third-degree polynomial, which is based on Fermat's approach to the calculus of polynomials. This method uses a rigorous concept of the derivative but avoids ε-δ arguments.

We show how to find the extrema of polynomials in an elementary way and, thereby, how to find the number of their roots. More precisely, we reduce the problem of finding extrema to the problem of finding roots.

To motivate the reader, we first give some results that can be obtained by analyzing extrema and finding roots (Theorems 7.1.8–7.1.11). This will show that it is possible to easily apply the results without delving into the method of discovering and proving them. At this point the reader will be interested in learning the method. Indeed, the main point of this subsection is the method itself and not its applications.

We will demonstrate the general method using specific simple arguments. It is more convenient for the reader who is not familiar with calculus to have a direct elementary formulation and proof of the result, rather than deriving the result from more general results preceded by unmotivated theory. If a reader is interested in generalizations, this approach will motivate them to study the theory and helps them to learn it.

Although we do not use the notion of derivative, our presentation illustrates this concept with full rigor but without ε-δ notation. Therefore, the following presentation can be useful in studying the fundamentals of calculus. Unfortunately it is not well known (compare with [**Pon84**]). For further development of Fermat's approach to polynomials, see sections 2 and 3. The development of the idea of "graphs of functions" is described in [**FT07**, **Gor10**, **Tab88**]. For example, this idea can be applied to *pqr*-lemmas as [**DMSF**].

The fascinating history of these discoveries is described, for example, in [**Yu70**].

1.B. Problems

It is known that the graph of any quadratic polynomial has an axis of symmetry.

7.1.1. (a) The graph of any cubic polynomial has a center of symmetry.

(b) Find the coordinates of the center of symmetry of the graph of the function $y = -2x^3 - 6x^2 + 4$.

(c) Is it true that the graph of any polynomial of fourth degree has an axis of symmetry?

It is known that the quadratic equation $ax^2 + bx + c = 0$ has two solutions when $D > 0$, one solution for $D = 0$, and no solutions for $D < 0$, where $D := b^2 - 4ac$. A method for finding the number of solutions of a cubic equation without actually solving the equation is easy to derive directly (see problem 7.1.4 below). In particular, to solve the following problems it is not necessary to know formulas for the roots of cubic equations. Moreover, solutions that do not use formulas for the roots of a cubic are easier to derive than the formulas themselves. Compare with problems 7.1.5 and 7.2.2 below.

Here one can use the Intermediate Value Theorem (7.1.13) without proof.

7.1.2. How many real solutions do the following equations have?
(a) $x^3 + 2x + 7 = 0$;   (b) $x^3 - 4x - 1 = 0$.

7.1.3. (a) The equation $x^3 + x + q = 0$ has exactly one solution for any q.

(b) Under what condition on p and q does the equation $x^3 + px + q = 0$ have exactly two solutions?

(c) Express these two solutions in terms of p and q.

7.1.4. (a) Find the intervals where $f(x) = x^3 - 6x + 2$ is increasing and where it is decreasing.

(b) For the same function find the maximum and minimum values on the interval $[0, 3]$.

(c) For which q will $x^3 - x + q = 0$ have exactly one solution?

(d) How can one determine the number of solutions to the equation $x^3 + px + q = 0$? Your answer should be in terms of p and q.

(e) Likewise, how can one determine the number of solutions to each of the equations $ax^3 + bx^2 + cx + d = 0$ in terms of a, b, c, and d?

7.1.5. How can one determine the number of solutions to the equations below? Your answers should be in terms of p and q.

(a) $x^4 + x + q = 0$; (b) $x^4 + px + q = 0$; (c) $x^n + px + q = 0$.

Hints

7.1.1. (a) First prove the statement for the trinomials $ax^3 + cx$ and $ax^3 + cx + d$.

7.1.2. (a) *Answer*: 1. Let $f(x) := x^3 + 2x + 7$. Since $f(-2) < 0$ and $f(1) > 0$, the Intermediate Value Theorem (7.1.13) implies that there is a root. Since f is monotone, there is only one root.

(b) *Answer*: 3. Let $f(x) := x^3 - 4x - 1$. Since $f(-2) < 0$, $f(-1) > 0$, $f(0) < 0$, and $f(3) > 0$, the Intermediate Value Theorem implies that there are three roots.

7.1.4. (a) Consider the sign of $\frac{f(x_1)-f(x_2)}{x_1-x_2}$.

(c) See (a) and (b). Find the intervals where the function increases and where it decreases. Find the points of local extrema and the function values at these points.

(d) Reduce to (c) by the substitution $y = kx$.

(e) Reduce to (d) using a substitution.

7.1.4. (d) *Answer*: If $p = q = 0$ then there is one root; otherwise let $D := \left(\frac{p}{3}\right)^3 + \left(\frac{q}{2}\right)^2$. If $D > 0$ there is one root, if $D = 0$ there are two roots, and if $D < 0$ there are three roots.

7.1.5. (b) *Answer*: If $p = q = 0$ then there is one root; otherwise let $D := \left(\frac{p}{4}\right)^4 + \left(\frac{q}{3}\right)^3$. If $D > 0$ there is one root, if $D = 0$ there are two roots, and if $D < 0$ there are three roots.

1.C. Statements of the main results

The following are standard facts from the school curriculum.

Theorem 7.1.6. Let a and b be real numbers. Then the following conditions are equivalent:

(1) There exist real numbers x and y such that $a = x + y$ and $b = xy$.

(2) The equation $t^2 - at + b = 0$ has a real root.

(3) $4b - a^2 \leq 0$.

Here is a closely related result: The quadratic equation $t^2 - at + b = 0$ has two roots for $D := a^2 - 4b > 0$, one root for $D = 0$, and no roots for $D < 0$.

A function f is called *strictly increasing* on an interval if $f(t_1) > f(t_2)$ for any $t_1 > t_2$ in this interval. *Strictly decreasing* functions are defined a similar way.

Theorem 7.1.7. Let a and b be real numbers. Then $t^2 - at + b$ is strictly decreasing on $(-\infty, a/2]$ and is strictly increasing on $[a/2, +\infty)$.

These theorems, as well as the formula for the roots of a quadratic equation, are proved using the equality

$$t^2 - at + b = \left(t - \frac{a}{2}\right)^2 + \left(b - \frac{a^2}{4}\right).$$

In this section, we explore the notion of a derivative by examining the well-known generalization of the above theorems to three numbers. (This generalization dates back to Fermat and possibly even earlier.) We will start with special cases.

Theorem 7.1.8. Let b and c be real numbers. Then the following conditions are equivalent:

(1) There exist real numbers x, y, and z such that $0 = x + y + z$, $b = xy + yz + zx$, and $c = xyz$.

(2) The equation $t^3 + bt - c = 0$ has three real roots, taking into account multiplicity.

(3) $4b^3 + 27c^2 \leq 0$.

Note that condition (3) obviously does not hold when b is greater than 0, and is equivalent to "$b \leq -3\sqrt[3]{c^2/4}$" or "$b \leq 0$ and $|c| \leq 2\sqrt{-b^3/27}$."

Theorem 7.1.9. Let b and c be real numbers, and let $f(t) := t^3 + bt - c$.
- If $b \geq 0$, then f is strictly increasing on $(-\infty, +\infty)$.
- If $b < 0$, then f is strictly increasing on $(-\infty, -\sqrt{-b/3}]$, is strictly decreasing on $[-\sqrt{-b/3}, \sqrt{-b/3}]$, and is strictly increasing on $[\sqrt{-b/3}, +\infty)$.

Now we give the formulation of the general case. It is more cumbersome but more useful. Also, it may not yet be obvious to the reader how the general case can be reduced to the special case.

Theorem 7.1.10. Let a, b, and c be real numbers. Then the following conditions are equivalent:

(1) There exist real numbers x, y, and z such that

$$a = x + y + z, \quad b = xy + yz + zx, \quad \text{and} \quad c = xyz.$$

(2) The equation $t^3 - at^2 + bt - c = 0$ has three real roots, taking into account multiplicity.

$$(3) \ 4\left(b - \frac{a^2}{3}\right)^3 + 27\left(c - \frac{ab}{3} + \frac{2a^3}{27}\right)^2 \leq 0.$$

This result is very useful. For example, [**DMSF**, problem 2] is a special case of an equivalent version of condition (3) of Theorem 7.1.10, similar to the version given after Theorem 7.1.8. For applications to elementary inequalities, see [**SB78**, problems 12 and 32] and [**Go09**].

Theorem 7.1.11. Let a, b, and c be real numbers, and let $f(t) := t^3 - at^2 + bt - c$.
- If $3b \geq a^2$, then f is strictly increasing on $(-\infty, +\infty)$.
- If $3b < a^2$, then f is strictly increasing on $\left(-\infty, \dfrac{a-\delta}{3}\right]$, is strictly

decreasing on $\left[\dfrac{a-\delta}{3}, \dfrac{a+\delta}{3}\right]$, and is strictly increasing on $\left[\dfrac{a+\delta}{3}, +\infty\right)$;
here $\delta = \sqrt{a^2 - 3b}$.

1.D. Proofs

Proof of the equivalence $(1) \Leftrightarrow (2)$ **in Theorems 7.1.8 and 7.1.10.**
By definition, the existence of three real roots (with multiplicity) of the equation $t^3 - at^2 + bt - c = 0$ means the existence of real numbers x, y, and z for which
$$t^3 - at^2 + bt - c = (t - x)(t - y)(t - z)$$

(that is, the coefficients at the corresponding degrees in the two polynomials are equal). This is equivalent to condition (1) by Vieta's Theorem 3.6.5 (by multiplying out the terms on the right-hand side). □

Reducing Theorem 7.1.10 to the special case of $a = 0$, that is to Theorem 7.1.8. Let $u := t - \frac{a}{3}$. Then $t = u + \frac{a}{3}$, so
$$t^3 - at^2 + bt - c = u^3 + \left(b - \frac{a^2}{3}\right)u - \left(c - \frac{ab}{3} + \frac{2a^3}{27}\right).$$

Therefore, Theorem 7.1.10 follows from the special case of $a = 0$, that is, from Theorem 7.1.8.

7.1.12. Reducing Theorem 7.1.11 to the special case of $a = 0$, that is, to Theorem 7.1.9.

Proof of Theorem 7.1.9 for $b \geq 0$. Since $b \geq 0$, the function $t^3 + bt - c$ is a sum of increasing functions (at least one of which is strictly increasing) and hence is strictly increasing. □

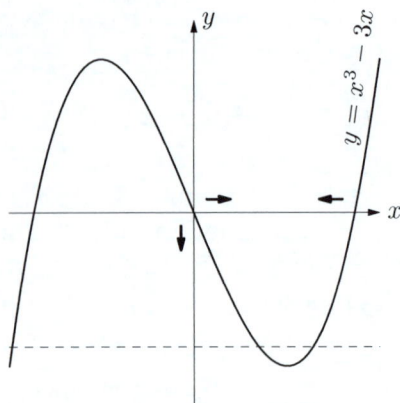

FIGURE 7.1. Graph of the function $f(t) = t^3 - 3t$

Proof of the equivalence $(2) \Leftrightarrow (3)$ **in Theorem 7.1.8 for** $b \geq 0$.
First, assume that the required roots exist. Since $b \geq 0$, the function $t^3 + bt - c$ is strictly increasing, as it is the sum of increasing functions, one of which is strictly increasing. Therefore the equation $t^3 + bt - c = 0$ has no more than one real root. This and the equation $x + y + z = 0$ imply $x = y = z = 0$. Therefore $4b^3 + 27c^2 = 0$. Now assume that $4b^3 + 27c^2 \leq 0$. Then $b = c = 0$, so we can take $x = y = z = 0$. $\square$

To show how to find intervals on which a function is increasing and decreasing for functions, we derive the equivalence $(2) \Leftrightarrow (3)$ in Theorem 7.1.8 for $b < 0$ from Theorem 7.1.9 for $b < 0$. After this, a simpler proof of this equivalence, due to M. Gorelov, is given. For yet another proof of Theorems 7.1.8 and 7.1.10, which involves using complex numbers and calculating the *discriminant* of a cubic polynomial in terms of its coefficients, see [**DMSF**, section 2, problems 6–22]. Although this proof is longer than each of these, it illustrates other important interesting ideas. Also, [**Tab88**] presents a geometric interpretation (but not a proof) of these theorems.

Heuristic considerations for the derivation of Theorems 7.1.9 *and* 7.1.8 (not formally used in the proof).

The previous proof shows that to derive Theorem 7.1.8 it is necessary to find out how many roots $f(t) := t^3 + bt - c$ has. And to do this you need to find the local maxima and minima of f, that is, to establish Theorem 7.1.9. The constant c doesn't affect this. Figure 7.1 shows the graphs of the function $f(t) = t^3 + bt$ for different b. It's clear that

- if $b \geq 0$, then f is increasing, and
- if $b < 0$, then f has a local maximum and a local minimum.

Let us show how to find the local maximum and local minimum for the example where $b = -3$. (The general case can be reduced to it, which is done below, or be derived in a similar fashion.) In other words, we will find the local maximum and minimum of the function $f(t) := t^3 - 3t$. The condition that f is strictly increasing is equivalent to the condition $\varphi(t_1, t_2) > 0$ for any distinct t_1 and t_2, where

$$(*) \qquad \varphi(t_1, t_2) := \frac{f(t_1) - f(t_2)}{t_1 - t_2} = t_1^2 + t_1 t_2 + t_2^2 - 3.$$

If these conditions are satisfied for "sufficiently close" t_1 and t_2, then, by transitivity, they are satisfied for all t_1 and t_2. In other words, we need to examine the values of t_1 and t_2 where ϕ changes sign. Thus we come to the conjecture that the boundary points of the intervals on which f is monotone are the roots of the equation $t^2 + tt + t^2 - 3 = 0$. These roots are equal to ± 1. (These arguments are similar to those in [**Ben88**]. Looking at a simple example before explaining the general method helps to make the method easier to understand.)

Proof of Theorem 7.1.9 for $b < 0$. We can assume that $c = 0$. Setting $u := t\sqrt{-b/3}$, we can assume that $b = -3$. Define $\varphi(t_1, t_2)$ by the formula $(*)$ for $f(t) := t^3 - 3t$. Then $\varphi(t_1, t_2) > 0$ for any distinct $t_1, t_2 \geq 1$. Consequently, $f(t)$ is strictly increasing on $[1, +\infty)$. The two other assertions of the theorem are proved similarly. $\qquad\qquad \square$

Proof of the equivalence $(2) \Leftrightarrow (3)$ in Theorem 7.1.8 for $b < 0$. Setting $u := t\sqrt{-b/3}$, we can assume that $b = -3$. Let $f(t) := t^3 - 3t - c$. We have $-f(t) = c - t(t^2 - 3)$, and therefore

$$f(-1)f(1) = (c + 2 \cdot 1)(c - 2 \cdot 1) = c^2 - 4 = (4b^3 + 27c^2)/27.$$

Thus, the inequality (3) is equivalent to the condition $f(-1)f(1) \leq 0$.

Suppose that the required roots x, y, and z exist. Since $-3 = b < 0$, the case of $x = y = z$ is impossible. Therefore the equation $f(t) = 0$ has at least two different real roots. Denote by t_+ the larger of the numbers 2 and $1 + |c|$. Then

$$t_+ \geq 1 + |c| > 1 \quad \text{and} \quad f(t_+) > (1 + |c|)(2^2 - 3) - c > 0.$$

Similarly, it is proved that there exists $t_- < -1$ such that $f(t_-) < 0$. Since the equation $f(t) = 0$ has at least two different real roots, using Theorem 7.1.9 for $b = -3$ we see that $f(-1)$ and $f(1)$ have different signs, that is, $f(-1)f(1) \leq 0$.

Now suppose that $f(-1)f(1) \leq 0$. If $f(-1)f(1) < 0$, then reversing the argument from the previous paragraph and using the Intermediate Value Theorem 7.1.13 (see below), we see that the equation $f(t) = 0$ has three real roots. We denote them by x, y, and z. If $f(-1)f(1) = 0$, then $c^2 - 4 = 0$. We set $x = y = -\operatorname{sgn} c$ and $z = 2\operatorname{sgn} c$, where $\operatorname{sgn} x = 1$ if $x > 0$, -1 if $x < 0$, and 0 if $x = 0$. (The reader probably guessed how to choose these formulas.) $\qquad\qquad \square$

Theorem 7.1.13 (Intermediate Value Theorem). Let f be a polynomial and let $a < b$. If $f(a) > 0 > f(b)$, then there exists $c \in (a, b)$ such that $f(c) = 0$.

Sketch of another proof of the equivalence (2) $\Leftrightarrow$ (3) **in Theorem 7.1.8.**(For the case $b < 0$ with an argument due to M. Gorelov.) First we repeat the first paragraph of the previous proof: Setting $u := t\sqrt{-b/3}$, we can assume $b = -3$. Let $f(t) := t^3 - 3t - c$.

Assume that the required roots x, y, and z exist. We assume that they are all different from ± 1 (this case can be considered separately). Then from the equality $-3 = xy + xz + yz = -x^2 - xy - y^2$ it follows that on each of the intervals $(-\infty, 1)$, $(-1, 1)$, and $(1, +\infty)$ there is at most one root. Therefore, each of the intervals contains exactly one root. Then from the equality $f(t) = (t - x)(t - y)(t - z)$ it follows that $f(-1) > 0$ and $f(1) < 0$. Therefore, $f(-1)f(1) < 0$.

Now suppose that $f(-1)f(1) \le 0$. Then, by the Intermediate Value Theorem, the equation $f(t) = 0$ has a root $x \in [-1, 1]$. Therefore we have $f(t) = f(t) - f(x) = (t - x)(t^2 + xt + (x^2 - 3))$. We assume that $x \ne \pm 1$ (this case can be considered separately). Since $x \in (-1, 1)$, the discriminant of the square trinomial $t^2 + xt + (x^2 - 3)$ (from t) is positive. Therefore, by Theorem 7.1.6, the equation $f(t) = 0$ has two more roots y and z. $\qquad\square$

7.1.14. (a) Prove Theorem 7.1.9.

(b) Find the largest intervals where f is strictly increasing or strictly decreasing for $f(t) = t^4 - 4t$ and $f(t) = t^4 - 12t^3 + 22t^2 - 24t + 10$.

7.1.15. Any polynomial of odd degree has a root.

2. Introductory analysis of polynomials (2)

For a finite sequence $b_0, \ldots, b_k$ of nonzero numbers, an index $i \in \{1, \ldots, k\}$ such that the numbers b_{i-1} and b_i have different signs is called the *change of sign*.

(For a finite sequence that contains zeros, its *change of sign* is the change of sign in the sequence of nonzero members obtained from the given sequence by removing all the zeros.)

7.2.1. (a) The number of *positive* solutions of $ax^2 + bx + c = 0$ does not exceed the number of changes of sign in the sequence a, b, c.

(b) The number of positive solutions of $ax^3 + bx^2 + cx + d = 0$ does not exceed the number of changes of sign in the sequence a, b, c, d.

7.2.2. (a) **Descartes' rule of signs.** The number of positive solutions of the equation $p_n x^n + \ldots + p_1 x + p_0 = 0$ does not exceed the number of sign changes in the sequence $p_0, \ldots, p_n$.

(b) Modify Descartes' rule of signs to estimate the number of negative roots of a given polynomial.

(c)* Modify Descartes' rule of signs to estimate the number of roots of a given polynomial in a given interval $[a, b]$.

(d) **The MacLaurin inequalities.** For $x_1, \ldots, x_n > 0$ define

$$M_k := \sqrt[k]{\frac{\sum\limits_{i_1 < \ldots < i_k} x_{i_1} \cdot \ldots \cdot x_{i_k}}{\binom{n}{k}}}.$$

Note that M_1 is the arithmetic mean and M_n is the geometric mean. Then

$$M_1 \geq M_2 \geq \ldots \geq M_n.$$

7.2.3. (a) For an even n, the polynomial $\sum\limits_{k=0}^{n} \frac{x^k}{k!}$ does not have real roots, and for an odd n it has exactly one real root.

(b) The minimum absolute value of the roots of the polynomial $\sum\limits_{k=0}^{n} \frac{x^k}{k!}$ tends to infinity as $n \to \infty$.

To solve these and many other problems, the following concept is needed.

The *pre-derivative* of a polynomial f is the polynomial $D_f(x, y) := \frac{f(y) - f(x)}{y - x}$ in two variables x and y. (Verify that this is indeed a polynomial.)

The *derivative* of f is the polynomial $f'(x) := D_f(x, x)$. A geometrical interpretation is that the equation of the tangent line to the graph $y = f(x)$ of a polynomial function f at the point $(a, f(a))$ is given by $y = f'(a)(x - a) + f(a)$. Formally, this can be understood as the definition of the tangent line.

7.2.4. Provide precise formulations of the following assertions, and prove them:

(a) $(f + g)' = f' + g'$;   (b) $(af)' = af'$;   (c) $(x^n)' = nx^{n-1}$;

(d) $(p_n x^n + \ldots + p_1 x + p_0)' = np_n x^{n-1} + (n-1)p_{n-1} x^{n-2} + \ldots + p_1$ (for $n = 0$ this expression is equal to 0).

(e) **Leibniz's rule.** $(fg)' = f'g + fg'$.

7.2.5. (a) **Lemma on the sign-preserving property of polynomials.** For any number a and polynomial g, if $g(a) > 0$ then there exists a positive $\delta = \delta(g, a)$ such that $g(x) > 0$ for any $x \in (a - \delta, a + \delta)$.

(b) **Fermat's Theorem.** If a is a point of local minimum or maximum of a polynomial f, then $f'(a) = 0$.

(c) Is the converse of (b) true?

(d) If a polynomial f is increasing (not necessarily strictly)[1] on an interval, then the derivative f' is non-negative on this interval.

(e) Is it true that if a polynomial f is strictly increasing on an interval, then its derivative f' is positive on this interval?

7.2.6. (a) If the derivative of a polynomial is positive on an interval, then the polynomial is strictly increasing on this interval.

(b) If the derivative of a polynomial is non-negative on an interval, then the polynomial is increasing (not necessarily strictly) on this interval.

7.2.7. (a) **Existence of extrema.** Every polynomial is bounded on any closed interval, and attains the maximum and minimum on this interval.

(b) **Rolle's Theorem.** Between any two roots of a polynomial lies the root of its derivative.

(c) **Lagrange's Mean Value Theorem.** For any $a \neq b$ and a polynomial f there exists $c \in [a, b]$ such that $f'(c) = \frac{f(a)-f(b)}{a-b}$.

Theorem 7.2.8 (Taylor's formula). (a) For any $a \neq b$ and polynomial f there exists $c \in [a, b]$ such that

$$f(b) = f(a) + \frac{f'(a)}{1!}(b - a) + \frac{f''(c)}{2!}(b - a)^2.$$

(b) For any $a \neq b$ and polynomial f of degree n there exists $c \in [a, b]$ such that

$$f(b) = f(a) + \frac{f'(a)}{1!}(b - a) + \frac{f''(a)}{2!}(b - a)^2 + \ldots$$
$$+ \frac{f^{(n-1)}(a)}{(n - 1)!}(b - a)^{n-1} + \frac{f^{(n)}(c)}{n!}(b - a)^n.$$

Hints

7.2.2. (a) Use induction on n and Rolle's Theorem 7.2.7 (b).

Another method. Denote by $d(f)$ the number of sign changes in the sequence of coefficients of the polynomial f. Prove that $d(f(x)(x - c)) \geq d(f(x)) + 1$ for any $c > 0$.

(c) Use Descartes' rule of signs and start with $b = \infty$.

7.2.4. (d) The statement follows from (a), (b), and (c).

---

[1] *Editor's note:* If a *non-constant polynomial* is increasing (decreasing) on an interval, it must be strictly increasing (decreasing). Constant functions are the only monotone but not strictly monotone polynomials on an interval.

7.2.5. (b) Assume to the contrary that $f'(a) \neq 0$. From (a) it follows that there exists δ such that $\frac{f(a+h)-f(a)}{h}$ and $f'(a)$ have the same sign for any $h \in (-\delta, \delta)$.

(c) No. A counterexample is provided by $f(x) = x^3$, $a = 0$.

7.2.6. (b) Apply Lagrange's Theorem 7.2.7 (c). There is also a direct proof which does not generalize to arbitrary continuous functions.

7.2.7. (a) Consider dividing the interval in half.

(b) See (a).

(c) See (b).

3. The number of roots of a polynomial (3*)

7.3.1. (a) If $p(x)$ is a polynomial with complex coefficients, the roots of $p'(x)$ lie inside the convex hull of the roots of $p(x)$. Compare with Rolle's Theorem 7.2.7 (b).

(b) The number of real roots of a polynomial p is equal to the number of vertical asymptotes of the graph of the function $y = \frac{p'(x)}{p(x)}$.

(c) Let p be a polynomial of degree n having n distinct real roots. Prove that the solution set of the inequality $\frac{p'(x)}{p(x)} > 1$ is the union of a finite number of intervals, and find the sum of their lengths.
(*International Mathematical Olympiad*, 1988).

(d) If $p(x) = (x - a_1) \cdots (x - a_n)$ then

$$\frac{p'(x)}{p(x)} = \frac{1}{x - a_1} + \ldots + \frac{1}{x - a_n}.$$

7.3.2.* Find the number of solutions to each equation (the answers will depend on p, q, r, s, t).

(a) $x^4 - x^2 + px + q = 0$; (b) $x^4 + x^2 + px + q = 0$;

(c) $x^4 + px^2 + qx + r = 0$; (d) $px^4 + qx^3 + rx^2 + sx + t = 0$.

The solutions of (c) and (d) show that finding the roots (without taking into account the multiplicity) of an arbitrary fourth-degree polynomial can be reduced to the cases (a) and (b). You should be able to work out specific cases, but you will most likely not be able to solve the problem in general without using the ideas below.

The following problems illustrate *Sturm's method* for finding the number of distinct real solutions (that is, the roots without multiplicities) of the equation $p_n x^n + \ldots + p_1 x + p_0 = 0$. Assertion 7.3.1 (b) suggests that it is necessary to find the number of asymptotes of the graph of a function p'/p without knowing the factorization of a polynomial p, i.e., its roots.

A point $x \in R$ is called a *point of ascent* of the function f if there exists $\varepsilon > 0$ such that $f(t) < f(x)$ for $x - \varepsilon < t < x$ and $f(t) > f(x)$ for $x < t < x + \varepsilon$.

A point $x \in R$ is called a *point of descent* of the function f if there exists $\varepsilon > 0$ such that $f(t) > f(x)$ for $x - \varepsilon < t < x$ and $f(t) < f(x)$ for $x < t < x + \varepsilon$.

For example, for $f(x) = x^2$ the point $x = 1$ is a point of ascent, $x = -1$ is a point of descent, and $x = 0$ is neither a point of ascent nor a point of descent.

The *algebraic number of preimages* of a value y of the function is defined to be $a(f, y) := u - d$, where u and d are the number of points of ascent and descent, respectively, in the preimage of the point y (that is, the set of x such that $f(x) = y$). It is understood that u and d must be finite.

For example, $a(x^2, 1) = 1 - 1 = 0$, $a(x^2, 0) = 0$, and $a(x^2, -1) = 0$.

7.3.3. The number of roots of a polynomial p is equal to $-a(p'/p, y)$ for sufficiently large y.

7.3.4. Find $a(f, y)$ for
   (a) $f(x) = x^3 - 3x + 1$ and $y = -1$;
   (b) $f(x) = x^3 - 3x + 1$ and $y = 100$;
   (c) $f(x) = x^3 - 3x + 1$ and $y = 0$;
   (d) $f(x) = x^3 - 3x + 1$ and $y = -100$;
   (e) $f(x) = x^4 + 2x^3 - x^2 + 4x + 1$ and $y = -100$;
   (f) $f(x) = 1/x$ and $y = 5$;
   (g) $f(x) = a_n x^n + \ldots + a_1 x + a_0$ and sufficiently large y;
   (h) $f(x) = a_n x^n + \ldots + a_1 x + a_0$ and arbitrary y.

7.3.5. (a) If p is a polynomial other than a constant, then for any a the number of solutions to $p(x) = a$ is finite.

(b) If $f = p/q$ is a non-constant rational function (that is, the ratio of polynomials p and $q \neq 0$), then for any a the number of solutions to $f(x) = a$ is finite.

(c) If a is a root of a polynomial p, then there exist an integer $k > 0$ and a polynomial g such that $p = (x - a)^k g$ and $g(a) \neq 0$.

7.3.6. Let $p = p_n x^n + \ldots + p_1 x + p_0$ and $q = q_m x^m + \ldots + q_1 x + q_0$ be polynomials without common non-constant factors and with $p_n q_m \neq 0$. Let $f := p/q$. If $n < m$ we additionally assume that $y \neq 0$, and if $n = m$ we assume that $y \neq p_n/q_m$. Then $a(f, y)$ does not depend on the choice of y; thus we will denote it by $a(f)$.

This fact can be used below.

7.3.7. Find $a(f)$ for
(a) $f(x) = \frac{1}{x}$;   (b) $f(x) = \frac{1}{x^3 - 3x + 1}$;   (c) $f(x) = x + \frac{1}{x}$;
(d) $f(x) = -x^2 + 4x + 1 + \frac{1}{x+2}$;   (e) $f(x) = \frac{x^3 - x^2 + 5}{x + 2}$;
(f) $f(x) = \frac{x+2}{x^3 - x^2 + 5}$.

7.3.8. Let g, p, and q be nonzero polynomials with $\deg p > \deg q$. Then
(a) $a(q/p) = -a(p/q)$;   (b) $a(g + \frac{q}{p}) = a(g) + a(\frac{q}{p})$.

7.3.9. Is it true that $a(f + g) = a(f) + a(g)$ for any rational functions f and g?

7.3.10. Construct an algorithm for finding the number of roots of the stated type of a given polynomial p:
(a) all;   (b) positive;   (c) on a given interval;   (d) counting multiplicity.

A famous unsolved problem asks how to find the number of complex roots of a polynomial (counting multiplicity) lying in the right half-plane.

Hints

7.3.1. (a, b, c). Use (d).

7.3.2. (c) Reduce to (a) and (b).
(d) Reduce to (c).

7.3.5. Use Bezout's Theorem and its corollaries.

Suggestions, solutions, and answers

7.3.1. In the notation of problem 7.3.5 (c), we have $\frac{p'(x)}{p(x)} = \frac{k}{x-a} + \frac{g'}{g}$.

7.3.5. (a) Suppose that $p(x) = a$ has infinitely many solutions. Then the polynomial $q(x) := p(x) - a$ has infinitely many roots, which means that $q(x)$ is identically equal to 0. Therefore $p(x)$ is identically equal to a.

7.3.6. Use the Intermediate Value Theorem (7.1.13), plus the fact that a non-constant rational function has a finite number of roots.

7.3.9. *Answer*: No.

7.3.10. (a) **Sturm's Theorem.** For a nonzero polynomial p, define nonzero polynomials $q_1, \ldots, q_k$ such that

$$\frac{p'}{p} = \frac{1}{q_1 + \frac{1}{\ldots + \frac{1}{q_k}}}.$$

For a polynomial $g(x) = g_n x^n + \ldots + g_1 x + g_0$ with $g_n \neq 0$, define

$$a(g) := \begin{cases} 0 & \text{for even } n, \\ 1 & \text{for odd } n \text{ and } g_n > 0, \\ -1 & \text{for odd } n \text{ and } g_n < 0. \end{cases}$$

The above definition of the number $a(g)$ is equivalent to that given in problem 7.3.6 in view of problem 7.3.4 (h).

The number $N(p)$ of solutions to equation $p(x) = 0$ is equal to

$$a(q_1) - a(q_2) + \ldots + (-1)^{k+1} a(q_k).$$

The given definition of the number $a(g)$ is equivalent to that given in problem 7.3.6 in view of the result of problem 7.3.4 (h).

(b) We can assume that $p(a) \neq 0$. Then the number $N_+(p)$ of positive roots is equal to $N_+(p) = N(p(x^2))/2 = -a(\frac{xp'(x^2)}{p(x^2)})/2$.

(c) The number of roots in the interval $[a, b]$ is equal to $N_+(p(x - a)) - N_+(p(x - b))$.

(d) The number of roots of a polynomial p of degree $n + 1$, counting multiplicity, is equal to

$$N(p) + N(\gcd(p, p')) + N(\gcd(p, p', p'')) + \ldots + N(\gcd(p, p', \ldots, p^{(n)})).$$

4. Estimations and inequalities (4*)
By V. A. Senderov

The concept of derivative used this section was defined for polynomials in section 2; for the general case see, for example, the book [**Zor15**].

7.4.1. Compare (a) e^π and π^e;   (b)* 2^π and π^2;   (c) $\log_3 4$ and $\log_4 5$;
(d) $\log_{n-1} n$ and $\log_n(n + 1)$ for $n > 2$;
(e) $\log_3 4 \cdot \log_3 6 \cdot \ldots \cdot \log_3 80$ and $2 \log_3 3 \cdot \log_3 5 \cdot \ldots \cdot \log_3 79$;
(f) $\log_3 5$ and $\log_4 6$;   (g) $10^{\sqrt{11}}$ and $11^{\sqrt{10}}$;
(h)* $6^{\sqrt{7}}$ and $7^{\sqrt{6}}$;   (i) $\cot \frac{5\pi}{18}$ and $\frac{5\pi}{18}$.

7.4.2. Prove the following inequalities.
(a) $x \cos x < 0.62$ for $0 < x < \pi/2$;
(b) $\sin(\pi/18) > 0.17$;
(c) $\left(\frac{\sin x}{x}\right)^3 > \cos x$ for $0 < x < \pi/2$;

(d)* $\frac{1}{\sin^2 x} \le \frac{1}{x^2} + 1 - \frac{4}{\pi^2}$ for $0 < x \le \pi/2$;

(e) $\cos^{\cos^2 x} x > \sin^{\sin^2 x} x$ and $\cos^{\cos^4 x} x < \sin^{\sin^4 x} x$ for $0 < x < \pi/4$;

(f) $2|\sin^n x - \cos^n x| \le 3|\sin^m x - \cos^m x|$ for $0 < x < \pi/2$ and positive integers n and m with $n > m$.

7.4.3. (a) Find all positive integer solutions to $x^y = y^x$.

(b) Find all real solutions to $x^y = y^x$.

(c)* For any integer $a > 0$, show that $x^y - y^x = a$ has a finite number of positive integer solutions.

Suggestions, solutions, and answers

7.4.1. (a) Taking logarithms of both sides, we see that it suffices to compare the numbers $\ln \pi/\pi$ and $\ln e/e$. These are values of the function $f(x) := \ln x/x$. Examining the derivative $f'(x) = (1 - \ln x)/x^2$, we see that this function increases on $(0, e]$ and decreases on $[e, +\infty)$. Thus $x = e$ is the global maximum. Consequently, $\ln \pi/\pi < \ln e/e$, so $\pi^e < e^\pi$.

(e) We have $\log_3 4 > \sqrt{\log_3 3 \cdot \log_3 5}, \ldots, \log_3 80 > \sqrt{\log_3 79 \cdot \log_3 81}$.

(i) Notice that $\frac{5\pi}{18} > \frac{17}{20}$.

7.4.2. (a) Note that $x \sin\left(\frac{\pi}{2} - x\right) < x\left(\frac{\pi}{2} - x\right) \le \frac{\pi^2}{16}$.

(b) Consider a cubic polynomial with integer coefficients, one of whose roots is $\sin(\pi/18)$.

(c, d) Rewrite the inequality as $x < \frac{\sin x}{\cos^{\frac{1}{3}} x}$, or $\varphi(x) := \sin x \cos^{-\frac{1}{3}} x - x > 0$. We have $\varphi'(x) > 0 \Leftrightarrow 2t^3 - 3t^2 + 1 > 0$, where $t = \cos^{\frac{2}{3}} x$. But $2t^3 - 3t^2 + 1 > 0$ for $0 < t < 1$.

5. Applications of compactness (4*)
By A. Ya. Kanel-Belov

This section contains harder problems with fewer hints. However, these are interesting problems on an important topic and, as far as we know, have not been published before for a general mathematical audience.

7.5.1. Warm-up problem involving a finite set. Start with a finite string of zeros and ones. Replace any "10" substring with "0001." Prove that eventually there will be nothing to replace, i.e., the process will end.

7.5.2. The idea of compactness. (a) Assume that humankind lives forever and the number of people in each generation is finite. Prove that there is an infinite male chain of descendants.

(b) In an infinite parliament, each member has no more than three ene-
mies. Prove that the parliament can be divided into chambers so that each
member will have no more than one enemy in their chamber.[2]

(c) It is known that any *finite* map on the plane can be properly colored
using 4 colors. Prove that an *arbitrary* (i.e., possibly infinite) map on the
plane can also be properly colored in 4 colors. (Countries can be regarded
as polygons. The coloring is called *proper* if any two countries sharing a
border are painted in distinct colors.)

7.5.3. For any M and k there is a sufficiently large v with the following
property: If all edges of a complete graph on v vertices are colored with M
colors, then there is a complete subgraph with k vertices, all of whose edges
are colored in one color.

7.5.4. From any infinite sequence of integers, it is possible to choose a
(infinite) subsequence such that each term is a multiple of the previous term
or no term is a multiple of any other term.

7.5.5. Consider an infinite set of points in the plane, no three of which lie
on a straight line. Then there is a convex figure whose boundary includes
infinitely many points from this set.

The ideas which are used to prove that certain algorithms eventually
stop often work along with the idea of compactness; indeed, these concepts
are related.

7.5.6. Does there exist an integer n such that any rational number between
0 and 1 can be represented as $\sum_{i=1}^{n} \frac{1}{a_i}$, where $0 < a_i \in \mathbb{Z}$?

7.5.7. (a) All finite sequences consisting of zeros and ones are partitioned
into two disjoint classes: *blue* and *red*. Prove that any infinite sequence of
zeros and ones can be split, perhaps omitting the first few members, into
finite pieces, all of the same color.

(b) An infinite sequence of digits are recorded on a tape. Prove that
either one can remove from it 10 hundred-digit numbers, in descending order,
or some combination of digits is repeated 10 times in a row.

7.5.8. Let $F \colon [0,1] \to [0,1]$ be an increasing continuous function. Prove

---

[2]The finite parliament case is analyzed in the next volume of this text (section 13.5,
"Semi-invariants"). *Challenge*: Which statements in this section are true for infinite sets,
and which are not?

that for any integer $N > 0$ the graph of the function can be covered with N rectangles whose sides are parallel to the coordinate axes, such that the area of each rectangle is equal to $1/N^2$.

7.5.9. Counterexamples in non-compact situations. Let $F(x)$ be a continuous function defined for $x \geq 0$.

(a) Suppose that $F(x+n) \to 0$ as $n \to \infty$ for each $x > 0$. Is it necessarily true that $F(x) \to 0$ as $x \to \infty$?

(b) Suppose that for any $x > 0$ the sequence $\{F(nx)\}$ converges. Is it necessarily true that the limit of $F(x)$ as $x \to \infty$ exists?

See [**AS16a**, section 5.2] for other applications of compactness.

Suggestions, solutions, and answers

7.5.2. (a, b) See (c).

(c) Call a proper coloring P of the set of countries S *infinitely extendable* if for any finite set S', P extends to $S \cup S'$. If the coloring is not infinitely extendable, then there exists a finite *set of nonextendability S'* such that P does not extend to $S \cup S'$.

If P is an infinitely extendable coloring of set S and C is a country, then there exists an infinitely extendable coloring P' of the set $S' = S \cup \{C\}$ that extends P. Indeed, there are at most four proper colorings of the set S' extending P. If each of them is nonextendable, then the union of the corresponding sets of nonextendability is the set of nonextendability for P.

The set of all countries is countable. Enumerate the countries as $C_1, \ldots, C_n, \ldots$. By hypothesis, the coloring of the empty set is infinitely extendable, and it can be sequentially extended to the infinitely extendable coloring of the first n countries for any n. The union of all such colorings will give the desired coloring for all the countries.

Remark. Using Zorn's lemma, it is possible to prove the following generalization: If any finite subgraph of a graph can be properly colored with k colors, then the entire graph can be properly colored with k colors.

Alternative solution (by B. Shoikhet). Enumerate the countries by integers. Enumerate the colors with the integers $1, 2, 3, 4$. Encode the coloring of the first n countries using the decimal fraction $0.a_1 a_2 \ldots a_n$ where a_i is the color of the ith country. Since the first n countries can be properly colored, there exists a sequence of numbers $\{x_n\} \subset [0,1]$ where x_n encodes the proper coloring of the first n countries. From any sequence of points in an interval, it is possible to extract a convergent sequence. Let X be a limit point of the sequence $\{x_n\}$, that is, the limit of some subsequence. It is easy to see that the decimal decomposition of X encodes a proper coloring of all countries.

7.5.8. In addition to the solution based on compactness, this problem has an easy solution by induction, where we use the Intermediate Value Theorem and consider transformations that are invariant with respect to dilations.

Chapter 8

Solving algebraic equations

Listeners are prepared to accept unstated (but hinted)
generalizations much more than they are able ...
to decode a precisely stated abstraction and to re-invent
the special cases that motivated it in the first place.

P. Halmos, *How to talk mathematics.*

1. Introduction and statement of results

1.A. What is this chapter about?

Famous theorems of Gauss, Ruffini, Abel, Galois, and Kronecker (8.1.5, 8.2.2, 8.1.12, 8.1.13, and 8.1.14), about the constructibility of regular polygons and the insolvability of algebraic equations in radicals are classical results of algebra which are also important in theoretical computer science.

The definitions of constructibility and solvability in radicals as well as the statements of these theorems are given in subsections 1.B–1.D. We do not give the history of these theorems but direct the interested reader to the texts [**Gin72**, **Gin76**, **Man63**].

The main goal of this chapter is an exposition of deep algebraic ideas (more precisely, of Galois theory) via simple and beautiful proofs of these theorems (see 1.E). Our intended audience is anyone who appreciates this type of exposition: high school and university students, teachers, and professional mathematicians. Remarkably, these proofs (subsection 2.E and section 4) only require being able to prove irrationality (Chapter 3, section 1), to divide polynomials with a remainder (Chapter 3, section 3 and problems 3.4.3 and 3.4.4), to find roots of a complex number (problem 3.5.4), to multiply permutations (Chapter 4, section 1), and to solve systems of linear equations. For each individual proof, only some of these tools are needed. In addition to the simplicity of these proofs, they also illustrate several fundamental ideas of algebra (specifically, Galois theory).

Studying these proofs (even the initial arguments) helps one to better understand notions of "irrationality," "polynomials," "complex numbers," "permutations," and "linear algebra." Even those who do not comprehend a complete proof of the main results can gain a thorough understanding of these topics and can even solve research problems (see 1.E, [**Edw09**, **Est**, **Akh**, **Kog**, **Saf**], and references therein).

Before proving the insolvability of algebraic equations, we consider a general method for their solution: Lagrange's resolvent method (section 2). Indeed, the key idea of Abel and Galois is that if an equation is solvable in radicals at all, then it is solvable by this method. This idea is formalized by the Galois criterion 8.2.8 (a) for the solvability of an equation. Lagrange's resolvents are used to construct algorithms to determine whether the equation is solvable in radicals and to express roots in radicals for solvable equations.

For practical purposes, approximation methods for computing trigonometric functions and solving equations are more useful than precise formulas. Besides, equations can be solved using transcendental functions (see Vieta's method in section 2 of Chapter 3 and in [**PS97**]; for further development of these ideas see, e.g., [**Sko10**]). However, the problem of solvability in radicals is interesting as a test problem in modern theories of symbolic computation and computational complexity.

The proofs presented here are not assumed to be new. However, our exposition contains much pedagogical novelty (see 1.E and 1.F). Unfortunately these arguments are not well known. As a consequence, it is little known that not only solving quadratic and cubic equations but also proving the indicated theorems is more economical by not constructing and then applying Galois theory (as is typically done in standard algebra textbooks, like [**Kho13**, **Kir**]), but by proceeding directly (see references in 1.F), while at the same time discovering and using the basic ideas of the theory.

Plan for the chapter. It is not necessary to read the sections sequentially. For example, one can begin not with section 1 but by solving the problems in sections 2 and 3, since most of them use the previous material only for motivation. Readers can choose the sequence of study most convenient for them (or omit some parts altogether) on the basis of the plan presented below.

Subsections 1.B–1.D contain formulations of the main results. The next three subsections of section 1 are independent of the rest of the chapter (i.e., they are not used in the rest of the chapter, so it is sufficient to read 1.B–1.D). Subsection 1.G discusses a reformulation of Gauss's Theorem (mentioned in 1.B).

Plans for sections 2–4 are given at the beginnings of these sections. Proofs of the main results are given in 2.C, 2.E, and 4. Formally, they do not depend on the problems leading to them (in sections 2 and 3).

Acknowledgments. We thank A. Ya. Belov, I. I. Bogdanov, G. R. Chelnokov, P. A. Dergach, A. S. Golovanov, A. L. Kanunnikov, V. A. Kleptsyn, P. V. Kozlov, G. A. Merzon, A. A. Pakharev, V. V. Prasolov, A. D. Rukhovich, L. M. Samoilov, L. E. Shabanov, V. V. Shuvalov, M. B. Skopenkov, E. B. Vinberg, V. V. Volkov, M. N. Vyalyi, and J. Zung for useful discussions.

This chapter is based on lectures at the Moscow "Olympic" school, the Summer Conference of the Tournament of Towns [**ABG+**, **ECG+**], and the "Mathematical seminar" and "Olympiads and Mathematics".

1.B. Constructibility (1)

Remark 8.1.1. It is known that
$$\cos\frac{2\pi}{3} = -\frac{1}{2}, \quad \cos\frac{2\pi}{4} = 0, \quad \cos\frac{2\pi}{5} = \frac{\sqrt{5}-1}{4}, \quad \cos\frac{2\pi}{6} = \frac{1}{2},$$
$$\cos\frac{2\pi}{8} = \frac{1}{\sqrt{2}}, \quad \cos\frac{2\pi}{10} = \frac{\sqrt{5}+1}{4}, \quad \cos\frac{2\pi}{12} = \frac{\sqrt{3}}{2}.$$

For which numbers n is $\cos\frac{2\pi}{n}$ expressible by a similar formula? That is, for which n can we compute $\cos\frac{2\pi}{n}$ with a calculator that has only the four arithmetic operations and the square root button?

A real number is called *real constructible* if it can be obtained, starting with the number 1, from additions, subtractions, multiplications, divisions by nonzero numbers, and taking square roots of positive numbers. In other words, a real constructible number can be obtained on the calculator starting with the display 1, according to Remark 8.1.1 above.

For example, the following real numbers are real constructible:
$$\sqrt[4]{2} = \sqrt{\sqrt{2}}, \quad \sqrt{2\sqrt{3}}, \quad \sqrt{2}+\sqrt{3}, \quad \sqrt{1+\sqrt{2}}, \quad 1+\sqrt{3-2\sqrt{2}}, \quad \frac{1}{1+\sqrt{2}}.$$

Additionally, the values in 8.1.1 and 8.1.3 are real constructible.

The real constructibility of a number is equivalent to its constructibility with compass and straightedge. Therefore, the following results solve the famous problems of antiquity about constructions with compass and straightedge. We discuss this equivalence in subsection 1.G; however, it will not be used later. The study of real constructibility is also important as a trial problem for modern theories of symbolic computation and computational complexity; see, e.g., [**Kog**].

Theorem 8.1.2. The number $\sqrt[3]{2}$ is not real constructible.

See the proof in subsection 4.D.

A more formal statement of the question in Remark 8.1.1 is: For which numbers n is $\cos\frac{2\pi}{n}$ real constructible?

8.1.3. The number $\cos\frac{2\pi}{n}$ is real constructible for $n = 15, 16, 20, 24, 60$.

Lemma 8.1.4 (On multiplication of real numbers). (a) If $\cos\frac{2\pi}{n}$ is real constructible, then $\cos\frac{\pi}{n}$ is also real constructible.

(b) If $\cos \frac{2\pi}{n}$ and $\cos \frac{2\pi}{m}$ are real constructible and n and m are relatively prime, then $\cos \frac{2\pi}{mn}$ is real constructible.

Theorem 8.1.5 (Gauss). The number $\cos \frac{2\pi}{n}$ is real constructible if and only if $n = 2^\alpha p_1 \cdots p_l$, where for $l \geq 0$ the factors $p_1, \ldots, p_l$ are distinct primes of the form $2^{2^s} + 1$.

The constructibility in the theorem is proved in subsections 2.C and 2.E (or in subsection 2.F), and non-constructibility is proved in subsection 4.D.

Strictly speaking, Gauss's Theorem does not give a full solution to the problem of real constructibility of $\cos \frac{2\pi}{n}$ since it is not known which numbers of the form $F_s := 2^{2^s} + 1$ are prime.[1] However, the theorem provides a fast algorithm for determining constructibility.

Gauss's Theorem implies the non-constructibility of the number $\cos \frac{2\pi}{9}$ (however, it is easier to prove this directly; see problem 8.3.14 (a)). This implies the following result showing the impossibility of trisection of an angle with compass and straightedge.

Theorem 8.1.6. There exists α (for example, $\alpha = 2\pi/3$) such that $\cos \alpha$ is real constructible but $\cos(\alpha/3)$ is not.

1.C. Insolvability in real radicals

A real number is called *expressible by radicals* if it can be obtained, starting with 1, by finitely many operations of addition, subtraction, multiplication, division by a nonzero rational number, and taking the nth root of a positive number, where n is a positive integer. In other words, a real number a is expressible in real radicals if some set containing this number can be obtained starting from the set $\{1\}$ and using the following operations. To a given set $M \subset \mathbb{R}$ containing numbers $x, y \in M$ one can append the numbers

$$x + y, \quad x - y, \quad xy, \quad x/y \text{ if } y \neq 0,$$

and $\sqrt[n]{x}$ for $x > 0$ and a positive integer n.

This definition can be reformulated in terms of a calculator similarly to Remark 8.1.1. In standard terms we say that the number lies in some real radical extension of the field $\mathbb{Q}$ if it is expressible in real radicals.

A number a is expressible in real radicals if and only if there exist
- positive integers $s, k_1, \ldots, k_s$;

---

[1] If F_s is prime, it is called a Fermat prime. As of 2019, the only known Fermat primes are $F_0 = 3$, $F_1 = 5$, $F_2 = 17$, $F_3 = 257$, and $F_4 = 65537$.

- real numbers $f_1, \ldots, f_s$ and polynomials $p_0, p_1, \ldots, p_s$ with rational coefficients of $0, 1, \ldots, s$ variables respectively such that

$$\begin{cases} f_1^{k_1} = p_0 \quad \text{(a constant)}, \\ f_2^{k_2} = p_1(f_1), \\ \ldots \\ f_s^{k_s} = p_{s-1}(f_1, \ldots, f_{s-1}), \\ a = p_s(f_1, \ldots, f_s). \end{cases}$$

Remark 8.1.7. (a) Any real root of a quadratic equation with rational coefficients is expressible in real radicals.

(b) The equation $x^3 + x + 1 = 0$ has exactly one real root and this root is expressible in real radicals (see section 2); see also problem 8.2.3 (c).

(c) The equation $x^4 + 4x - 1 = 0$ has two real roots, both of which are expressible in real radicals (problem 3.2.6 (b)); see also problem 8.2.5 (d).

(d) Any real constructible number (1.B) is expressible in real radicals.

(e) There exists a cubic polynomial with rational coefficients, none of whose roots is expressible in real radicals (for example, $x^3 - 3x + 1$; this is proven in part (f) below.)

(f) The number $\cos(2\pi/9)$ is not expressible in real radicals.

Indeed, apply the triple-angle formula 3.1.5 (e) for cosine. We see that the numbers $\cos(2\pi/9)$, $\cos(8\pi/9)$, and $\cos(14\pi/9)$ are the roots of the equation $8y^3 - 6y + 1 = 0$. By Theorem 8.1.8 none of these numbers is expressible in real radicals.

(g) The trisection of an angle is impossible in real radicals. That is, there exists a number α (for example, $\alpha = 2\pi/3$) such that the number $\cos\alpha$ is expressible in real radicals and the number $\cos(\alpha/3)$ is not expressible in real radicals. (This follows from part (f).)

Theorem 8.1.8 (Solvability in real radicals)**.** For a cubic polynomial with rational coefficients the following conditions are equivalent:

(i) the polynomial has either at least one rational root or exactly one real root;

(ii) the polynomial has a root which is expressible in real radicals;

(iii) all real roots of the polynomial are expressible in real radicals.

The uniqueness of the real root of the "shortened" equation $x^3 + px + q = 0$ is equivalent to the following condition: $p = q = 0$ or $(p/3)^3 + (q/2)^2 > 0$; see problem 7.1.4 (d).

Clearly, (ii)⇔(iii) (this follows from Remark 8.1.7 (a)). The solvability in Theorem 8.1.8 (i.e., (i) ⇒ (ii)) can be proved by *del Ferro method* (see theorems given in the hints for problems 3.2.4 and 7.1.4 (d)); for another proof see 2.B.

The insolvability in Theorem 8.1.8 (i.e., (ii) ⇒ (i)) has a more complicated proof, see 4.E.

It is easier to prove the analogous result on *insolvability in polynomials*; see 3.F and 4.B.

Remark 8.1.9. From the insolvability in Theorem 8.1.8, it follows easily that *for any $n \geq 3$ there exists a polynomial of degree n, one of whose roots is not expressible in real radicals*. It is more difficult to prove an analogue of this statement with the words "one of the roots" replaced by "none of the roots"; see Theorem 8.1.10 below. At the same time, the roots of *some* equations of high degrees (for example, $x^5 = 2$) may well be expressible in real radicals (see 2.E).

A polynomial with coefficients in a set F is called *irreducible* over F if it cannot be decomposed into a product of polynomials of smaller degree with coefficients in F.

Theorem 8.1.10. If a polynomial of prime odd degree with rational coefficients is irreducible over $\mathbb{Q}$ and has more than one real root, then none of its roots are expressible in real radicals.

This is a real analogue of Kronecker's Theorem 8.1.14. The proof is given in 4.H.

Conjecture 8.1.11.* (a) Every real root of a polynomial of fourth degree with rational coefficients that is irreducible over $\mathbb{Q}$ is expressible in real radicals if and only if at least one root of its cubic resolvent (defined after problem 3.2.6 (b) is expressible in real radicals (cf. problem 8.3.13 (d)).

(b) If $\cos \frac{2\pi}{n}$ is expressible in real radicals, then it is real constructible (cf. Gauss's Theorem 8.1.5 on constructibility of regular polygons).

Perhaps the validity of these conjectures is known to specialists. Conjecture 8.1.11 (b) (and the answer to problem 8.3.3 with a proof sketch) was communicated by A. A. Kanunnikov. The reader may try to prove these conjectures after studying sections 3 and 4.

1.D. Insolvability in complex radicals (2)

Now we consider equations involving complex numbers. It turns out that a cubic equation (for example, $x^3 - 3x + 1 = 0$) that is not solvable in real radicals can be solved in complex radicals.

A complex number z is called *expressible in radicals* if it can be obtained, starting with 1, by finitely many operations of addition, subtraction, multiplication, division by a nonzero number, and taking the nth root, where n is a positive integer. In other words, a complex number is expressible in radicals if some set containing this number can be obtained starting from the set $\{1\}$ and using the following operations.

To a given set $M \subset \mathbb{C}$ containing numbers $x, y \in M$ one can add

$$x + y, \quad x - y, \quad xy, \quad x/y \text{ if } y \neq 0,$$

and any number $r \in \mathbb{C}$ such that $r^n = x$ for some integer $n > 0$.

This definition can be reformulated in terms of calculators similar to Remark 8.1.1. True, the calculator will be unusual: it works with complex numbers and, when the $\sqrt[n]{\ }$ button is pressed, outputs all n nth roots. In conventional terminology, we say that z lies in a radical extension of the field $\mathbb{Q}$.

For example, any (complex) root of a quadratic equation with rational coefficients is expressible in radicals. Similar assertions hold for equations of the third and fourth degrees. (These assertions can be proved by the *del Ferro and Ferrari methods*; see the theorems given in the hints for problems 3.2.4 and 3.2.7; for another proof see 2.B.[2]) However, similar assertions for equations of higher degrees do not hold.

Theorem 8.1.12 (Galois). There exists a fifth-degree polynomial with rational coefficients (for example, $x^5 - 4x + 2$), none of whose roots is expressible in radicals.

The famous problem of solvability in radicals was solved by the weaker Ruffini–Abel theorems that were proved a little earlier. Ruffini's Theorem 8.2.2 has a more complicated statement, but it leads to the proof of the Galois Theorem. The precise statement of Abel's Theorem is even more complicated and is not presented here (cf. [**Skod**, Remark 7]). An easier way to study the solvability problem is to prove (in 4.F) Theorem 8.1.13. This theorem is weaker than the Galois Theorem 8.1.12 and has a simpler proof. For $X \subset \mathbb{C}$, a complex number a is called *X-expressible in radicals* if a can be computed, starting with the set $X \cup \{1\}$, with finitely many operations of addition, subtraction, multiplication, division, and taking nth roots.

Theorem 8.1.13. There exist $a_0, a_1, a_2, a_3, a_4 \in \mathbb{C}$ such that no root of the equation $x^5 + a_4 x^4 + \ldots + a_1 x + a_0 = 0$ is $\{1, a_0, a_1, a_2, a_3, a_4\}$-expressible in radicals.

A similar result (with similar proof) holds for equations of any degree $n \geq 5$. The stronger Galois Theorem 8.1.12 is a consequence of the following result.

Theorem 8.1.14 (Kronecker). If a polynomial with rational coefficients is irreducible over $\mathbb{Q}$, has prime degree, has more than one real root, and has at least one non-real root, then the polynomial has no roots expressible in radicals.

---

[2]For an estimate on the number of required roots, see 3.I, 3.A, and [**ABG+**].

This theorem is interesting and nontrivial even for polynomials of degree 5. It is proved in 4.G. For the proof, the following generalization of Gauss's Theorem 8.1.5 is needed. Let

$$\varepsilon_q := \cos(2\pi/q) + i\sin(2\pi/q).$$

Theorem 8.1.15 (Gauss; lowering degree). (a) If q is a prime, then the number ε_q can be expressed in radicals using only roots of degree $q-1$.

(b) For every q the number ε_q can be expressed in radicals using only rth roots, with $r < q$.

Part (a) is proved similarly to the proof of constructibility in Gauss's Theorem (2.E and 2.F). Part (b) follows from (a) by induction on q. (The induction base is obvious. If $q = ab$ for some integers a and b with $0 < a, b < q$, then the inductive step follows from $\varepsilon_q = \sqrt[a]{\varepsilon_b}$. If q is a prime, then the inductive step follows by (a).)

The complex analogue of Remark 8.1.9 for $n \geq 5$ and Kronecker's Theorem 8.1.14 instead of Theorem 8.1.10 is valid. Moreover, the proof of Theorem 8.1.13 can be easily adapted to equations of any degree $n \geq 5$.

Theorem 8.1.16. For $a_{n-1}, \ldots, a_0 \in \mathbb{Q}$ there exists an algorithm for deciding whether all roots of the equation $x^n + a_{n-1}x^{n-1} + \ldots + a_1 x + a_0 = 0$ are expressible in radicals.

Theorem 8.1.16 can be proved using the Galois solvability criterion 8.2.8 (a) and an estimate 8.3.53 (b) of the number of operations.

1.E. What is special about our proofs

The proofs given here are much simpler and shorter than those presented in standard algebra textbooks. Here we mean a proof from scratch, and not a derivation of the result using previously developed theory. A comparison with proofs from less standard popular literature is given in 1.F.

This simplicity is due to the fact that, unlike most textbooks, the proofs given here do not use the term "Galois group" or even the term "group." Despite the absence of these *terms*, the *ideas* of the given proofs are the *starting point* for Galois theory and *constructive Galois theory* [**Edw09**].

Our proof of solvability is based on the *Lagrange resolvent* method. The proofs of insolvability use ideas of *symmetry* and *conjugation*. (A more formal description of the latter is the idea of an *automorphism of a field*; cf. [**Vag80**] for a wonderful exposition.)

The main ideas are presented via "olympiad" examples using the simplest special cases, free from technical details, and keeping terminology to a minimum. Although the main results concern equations of higher degree, our ideas are demonstrated using quadratic and cubic equations. Insolvability is proved initially under the condition that *the root was extracted only once* (in 3.A, 3.C, and 3.D). Consequently, the key examples involve rational numbers (not arbitrary fields or even field extensions over the rationals). These basic ideas (conjugations, fields, and others) are contained in lemmas about calculators, linear independence, and conjugation (sections 3 and 4). To prove Gauss's insolvability theorem 8.1.5, the degree of the *polynomial* is used (instead of the degree of the *field extension*). Before proving insolvability (8.1.12, 8.1.13, and 8.1.14) we prove insolvability in polynomials (Ruffini's theorems 8.2.2 and 8.4.4), as well as insolvability by *real* radicals of cubic equations (Theorem 8.1.8). Important ideas of proofs are explicitly emphasized as lemmas that are clearly formulated in simple language (on preservation of even symmetry (8.3.43 and 8.4.5), on powers of 2 (8.4.7), and on rationalization (8.4.13)).

This makes the proofs of insolvability more accessible by introducing interesting, clearly specified intermediate steps. In addition, this leads the reader to the assumption that the arguments here can be developed into a theory (Galois theory!), with many applications.

We show *how one can find* the presented proofs. Approaches to them are outlined in the form of problems in 2.D and 3. For the tradition of studying material via problems, see p. xvi. Although it is not easy to *find* proofs, it is possible to *present* them succinctly (see 2.E and 4). Skipping technical details is an important part of verification of the proof.

Many of these problems are good research topics for high school and junior university students in algebra, combinatorics, and computer science; see section 4. Examples of students' papers can be found in [**Saf, Akh, Kog**]. Good research problems include 8.2.7, 8.2.8, 8.3.3, 8.1.8, 8.1.10, 8.2.7, 8.2.11, 8.3.5, 8.3.9 (h), 8.3.17 (e), 8.3.19 (b), 8.3.21 (d), 8.3.25, 8.3.28, 8.3.32, 8.3.33, 8.3.34, 8.3.37, 8.2.2, 8.3.40, 8.3.45, and 8.3.51–8.3.53.

1.F. Historical comments

The proof of constructibility in Gauss's Theorem 8.1.5 is obtained from [**Edw97**, Ch. 24] after some simplification (we circumvent the use of Lemma 2; see the paragraph before problem 8.2.13 for details). It is simpler than the proof in [**KS08**]. An elementary proof of constructibility for $n = 17$ is given, for example, in [**BK13, Che34, Gin72, Pra07a, Pos14, PS97, Kol01**] and in [**Dör13**, Ch. 37] (wherein sometimes explicit formulas are given, either with proof of the assertions about the signs in front of the radicals, as in [**Dör13**, Ch. 37] and [**Saf**], or without a proof [**BK13**]). The general approach is outlined in [**Gau, Gin72**], where the clarity of proof is hampered a little by exposition of a general theory instead of proving a concrete result. The approach in [**Kir77**] provides an answer to the question of "why," and it would be interesting to develop it into a full proof.

The proof of non-constructibility in Gauss's Theorem 8.1.5 is similar to [**Dör13**, Supplement to Chapters 35–37]. It is simpler than the proof in [**KS08**].

We do not know whether a short direct proof of Theorems 8.1.8 and 8.1.10 on insolvability in real radicals has been published. The proof of Ruffini's Theorem (8.2.2 and 8.4.4) follows the proof given in the excellent book [**Kol01**]. We could not make out the proof presented there until we rediscovered it, explicitly stating the lemma on the preservation of even symmetry (8.3.43 and 8.4.5). The proof of Theorem 8.1.13 follows the proof given in [**PS97**]. We were able to verify the correctness of the ideas proposed there only after rewriting the proof and explicitly stating the rationalization lemma 8.4.13. Another proof of Abel's Theorem is given in [**Ale04**], [**FT07**, Lecture 5], and [**Sko11**].[3] The proof of Kronecker's Theorem 8.1.14 is based on the remarkable article [**Tik03**] and on the books [**Dör13**, Ch. 25] and [**Pra07a**, Supplement 8] (here the inaccuracies are corrected; see footnotes 13 and 16).

Other elementary expositions are given, for example, in [**Ber10**, **Bro**, **Had78**, **Vin80**, **Kan**, **L**, **Pes04**, **Ros95**, **Sti94**]. Note that the proofs in some of these sources are incomplete; see [**Skod**, Discussion].

The above elementary expositions were more useful to us (in spite of the drawbacks mentioned) than formal expositions (in standard textbooks presenting the theory) which start with several hundreds of pages of definitions and results whose role in the proof of the insolvability theorem is not clear at the point of their statements.

1.G. Constructions with compass and straightedge (1)

8.1.17. (a) Prove that starting with segments of lengths x and y it is possible to construct a segment of length $\sqrt{3xy + y\sqrt[4]{xy^3}}$ with compass and straightedge.

(b) Prove that starting with segments of lengths a, b, and c it is possible to construct segments of length $a + b$, $a - b$, ab/c, and $\sqrt{ab}$.

It follows from 8.1.17 (b) that if a segment of length 1 is given on a plane, then a segment of a real constructible length can be constructed with compass and straightedge. This simple result was already known to ancient Greeks. It turns out that the converse is also true.

Theorem 8.1.18 (Fundamental theorem on constructibility). If a segment of length a can be constructed with compass and straightedge, then the number a is real constructible.

---

[3]The proof in [**Ale04**] is presented in a shorter and simpler way in [**FT07**, Lecture 5] and [**Sko11**]. A large part of [**Ale04**] contains theory not required to prove the Abel–Ruffini Theorem. However, the author of [**Ale04**] succeeded in avoiding unmotivated exposition of the most complicated part of the theory.

This simple result (which was proved only in the 19th century) shows that the non-constructibility of the number $\cos(2\pi/n)$ implies the non-constructibility of a regular n-gon with compass and straightedge. To prove this result, we can consider all possible cases of the appearance of new objects (points, lines, circles) and show that the coordinates of all points constructed and the coefficients of the equations of all lines and circles drawn are real constructible. The reader will be able to complete the details independently or find them in [**Kol01**, **CR96**, **Man63**, **Pra07a**].

Hints

8.1.17. (a) It suffices, by applying (b), to construct segments of lengths $z_1 = \sqrt{xy}$, $z_2 = \sqrt{yz_1}$, $z_3 = 3x + z_2$, and $z = \sqrt{yz_3}$.

(b) Here is how to build a segment of length $\sqrt{ab}$. The height of a right triangle dropped to the hypotenuse is the geometric mean of the lengths of the segments into which it divides the hypotenuse. Therefore, if we are given segments with lengths a and b, by constructing a semicircle with diameter $a + b$ and finding its intersection with a straight line perpendicular to the diameter and dividing the diameter into segments of length a and b, we get a segment of length $\sqrt{ab}$.

2. Solving equations: Lagrange's resolvent method

We will demonstrate Lagrange's resolvent method with the simplest examples in 2.B. Its application to the proof of Gauss's constructibility theorem 8.1.5 is illustrated with examples and problems in 2.D. Constructibility is proved in 2.C and 2.E. In 2.C we prove a simpler part of the proof which does not use Lagrange resolvents. This section does not depend on previous ones. Material from 2.F is not used further.

See 1.B and 1.D for the definitions of real constructibility and expressibility in radicals. In this section equality signs involving a polynomial f (or f_j) mean equality of polynomials coefficientwise. Recall the notation

$$\varepsilon_q := \cos(2\pi/q) + i\sin(2\pi/q).$$

2.A. Definition of expressibility in radicals of a polynomial (1)

Let a and b be the roots of the quadratic equation $x^2 - (a + b)x + ab = 0$. The formulas

$$(a - b)^2 = (a + b)^2 - 4ab \quad \text{and} \quad a = \frac{a + b + (a - b)}{2}$$

show that *a root, a, of a quadratic equation is expressible in radicals* using the coefficients $a + b$ and ab of the equation. A rigorous definition of expressibility in radicals is given below.

Denote the elementary symmetric polynomials by

$$\sigma_1(x_1, \ldots, x_n) := x_1 + \ldots + x_n, \quad \ldots, \quad \sigma_n(x_1, \ldots, x_n) = x_1 \cdot \ldots \cdot x_n.$$

If the number n and the arguments $x_1, \ldots, x_n$ are clear from the context, we omit them from the notation.

A polynomial $p \in \mathbb{C}[x_1, \ldots, x_n]$ is called *expressible by* (complex) *radicals* if one can add p to the collection $\{\sigma_1, \ldots, \sigma_n\} \cup \mathbb{C}$ of polynomials by a sequence of the following operations:

• add the sum or the product of polynomials which are already in the collection;

• if some polynomial in the collection equals f^k for some $f \in \mathbb{C}[x_1, \ldots, x_n]$ and some integer $k > 1$, then add f to the collection.

Remark 8.2.1. (a) For example, if a collection contains $x^2 + 2y$ and $x - y^3$, then one may apply operations of the first type and add to the collection the polynomial
$$-5(x^2 + 2y)^2 + 3(x^2 + 2y)(x - y^3)^6.$$
If a collection already contains $x^2 - 2xy + y^2$, then one may apply the operation of the second type and add $x - y$ (or $y - x$).

(b) If we use only operations of the first type above, we can construct any polynomial with complex coefficients using polynomials which are already available.

(c) By Vieta's Theorem 3.6.5, $\sigma_1, \ldots, \sigma_n$ are the coefficients of the polynomial
$$t^n - \sigma_1 t^{n-1} + \ldots + (-1)^{n-1}\sigma_{n-1}t + (-1)^n \sigma_n \in \mathbb{C}[x_1, \ldots, x_n][t]$$
with roots $x_1, \ldots, x_n$. Therefore, the expressibility in radicals of the polynomial x_1 is equivalent to the expressibility (in the above sense) of its root x_1 in terms of the coefficients of this polynomial.

(d) The polynomial x_1 is expressible in radicals if and only if there exist

• positive integers $s, k_1, \ldots, k_s$;

• polynomials $f_1, \ldots, f_s$ in n variables and polynomials $p_0, p_1, \ldots, p_s$ in $n, n+1, \ldots, n+s$ variables respectively, with complex coefficients such that
$$\begin{cases} f_1^{k_1} = p_0(\sigma_1, \ldots, \sigma_n), \\ f_2^{k_2} = p_1(\sigma_1, \ldots, \sigma_n, f_1), \\ \ldots \\ f_s^{k_s} = p_{s-1}(\sigma_1, \ldots, \sigma_n, f_1, \ldots, f_{s-1}), \\ x_1 = p_s(\sigma_1, \ldots, \sigma_n, f_1, \ldots, f_s). \end{cases}$$
Here we omit the variables $(x_1, \ldots, x_n)$ in the polynomials $\sigma_1, \ldots, sigma_n$ and $f_1, \ldots, f_s$.

Theorem 8.2.2 (Ruffini). For every integer $n \geq 5$ the polynomial x_1 is not expressible by radicals.

The proof shows that, in fact, the polynomial $x_1x_2 + x_2x_3 + x_3x_4 + x_4x_5 + x_5x_1$ is not expressible in radicals for $n = 5$.

2.B. Solution of equations of low degrees (2)

8.2.3. Which of the following polynomials are expressible in radicals for $n = 3$?

(a) $(x - y)(y - z)(z - x)$;   (b) $x^9 y + y^9 z + z^9 x$;   (c) x.

To solve problem 8.2.3 and the following problems, one can use the Fundamental Theorem on Symmetric Polynomials 3.6.3. Hints for part (c) are problems 8.2.4 (a) and 8.2.6 (c).

8.2.4. A polynomial $f \in \mathbb{C}[u_1, u_2, \ldots, u_n]$ is called *cyclically symmetric* if $f(u_1, u_2, \ldots, u_n) = f(u_2, u_3, \ldots, u_{n-1}, u_n, u_1)$.

(a) Find at least one pair of numbers $\alpha, \beta \in \mathbb{C}$ such that the polynomial $(u + v\alpha + w\beta)^3$ is cyclically symmetric but the polynomial $u + v\alpha + w\beta$ is not.

(b) Express $x_1 x_3 + x_3 x_5 + x_5 x_7 + x_7 x_9 + x_9 x_1$ with finitely many applications of addition, subtraction, mulitiplication, division, and extracting roots, starting with several *cyclically symmetric* polynomials in $x_1, x_2, \ldots, x_{10}$.

8.2.5. Which of the following polynomials are expressible in radicals for $n = 4$?

(a) $(x - y)(x - z)(x - t)(y - z)(y - t)(z - t)$;
(b) $xy + zt$;   (c) $x + y - z - t$;   (d) x.

8.2.6. Solve the following systems of equations (x, y, z, t are unknowns; a, b, c, d are given constants):

(a) $\begin{cases} x + y + z + t = a, \\ x + y - z - t = b, \\ x - y + z - t = c, \\ x - y - z + t = d; \end{cases}$ (b) $\begin{cases} x + y + z + t = a, \\ x + iy - z - it = b, \\ x - y + z - t = c, \\ x - iy - z + it = d; \end{cases}$

(c) $\begin{cases} x + y + z = a, \\ x + \varepsilon_3 y + \varepsilon_3^2 z = b, \\ x + \varepsilon_3^2 y + \varepsilon_3 z = c. \end{cases}$

The expressions in problem 8.2.6 are called *Lagrange resolvents*. They are "better" than roots because they are "more symmetric" in the following sense.

Solution of a cubic equation using Lagrange resolvents (solution of problem 8.2.3 (c)). To find the roots for x, y, and z of a cubic equation, it suffices to find expressions a, b, and c from problem 8.2.6 (c). Notice that the del Ferro method from problem 3.2.2 leads us to the same expressions. By Vieta's Theorem 3.6.5, $a = a(x, y, z)$ is a coefficient of the equation. Under the substitution $x \leftrightarrow y$, the polynomial $b = b(x, y, z)$ goes to $\varepsilon_3 c$, and $c = c(x, y, z)$ goes to $\varepsilon_3^2 b$ (check this!). Therefore, the polynomials bc and $b^3 + c^3$ are invariant under this substitution. Similarly, they are invariant with respect to the substitution $z \leftrightarrow y$. Therefore the polynomials bc and $b^3 + c^3$ are *symmetric*, i.e., they do not change under *any* permutation of variables. From the Fundamental Theorem on Symmetric Polynomials (see, e.g., 3.6.3 (d)) and Vieta's Theorem 3.6.5 we see that the polynomials bc and $b^3 + c^3$ in x, y, and z can be represented as polynomials in the coefficients of the equation. Hence we can obtain b^3 and c^3 by solving certain quadratic equations. After that we obtain b and c.

Solution of an equation of fourth degree using Lagrange resolvents (solution to problem 8.2.5 (d)). To find the roots x, y, z, and t of an equation of degree four, it is enough to find expressions for a, b, c, and d from problem 8.2.6 (a). By Vieta's Theorem 3.6.5, a is a coefficient of the fourth-degree equation. The transposition $x \leftrightarrow y$ interchanges the polynomials c^2 and d^2 but does not change b^2. The cyclic permutation $x \to y \to z \to t \to x$ interchanges the polynomials b^2 and d^2 but does not change c^2. Therefore the polynomials b^2, c^2, and d^2 are permuted for every permutation of the variables x, y, z, and t. Hence their Vieta polynomials, i.e.,

$$b^2 + c^2 + d^2, \quad b^2 c^2 + b^2 d^2 + c^2 d^2, \quad b^2 c^2 d^2,$$

are symmetric. Consequently, these polynomials (in x, y, and z) can be represented as polynomials in the coefficients of the equation. Finally, by solving a cubic equation, we can get b^2, c^2, and d^2. Then it is easy to obtain b, c, and d.

Ruffini's Theorem 8.2.2 shows that the Lagrange resolvent method presented above for solving equations of degrees 3 and 4 (problems 8.2.3 (c) and 8.2.5 (d)) does not work for degree 5. Guess why!

Denote by Σ_q the set of permutations of the set $\{1, 2, \ldots, q\}$. For a permutation $\alpha \in \Sigma_q$ write

$$\vec{u}_\alpha := (u_{\alpha(1)}, \ldots, u_{\alpha(q)}).$$

Define the *Lagrange resolvent* by

$$t(u_1, \ldots, u_q) := \varepsilon_q u_1 + \varepsilon_q^2 u_2 + \ldots + \varepsilon_q^q u_q.$$

Define the *Galois resolvent* by

$$Q(u_1, \ldots, u_q, y) := \prod_{\alpha \in \Sigma_q} (y - t(\vec{u}_\alpha)) \in \mathbb{Q}[\varepsilon_q][u_1, \ldots, u_q, y].$$

8.2.7. (a) We have $Q(\varepsilon_q u_1, \ldots, \varepsilon_q u_q, y) = Q(u_1, \ldots, u_q, y)$.

(b) For some $R_Q \in \mathbb{Q}[\varepsilon_q][u_1, \ldots, u_q, z]$, $Q(u_1, \ldots, u_q, y) = R_Q(u_1, \ldots, u_q, y^q)$.

(c) If $x_1, \ldots, x_q \in \mathbb{C}$ are the roots of a polynomial $f \in \mathbb{Q}[x]$ of degree q, then $Q(x_1, \ldots, x_q, y) \in \mathbb{Q}[\varepsilon_q][y]$ and even $Q(x_1, \ldots, x_q, y) \in \mathbb{Q}[y]$.

The polynomial $R_Q(x_1, \ldots, x_q, z) \in \mathbb{Q}[z]$ is called *the resolvent polynomial* for f.

(d)* All the roots of the resolvent polynomial for $f(x) = x^5 + 15x + 44$ (and, therefore, all the roots of f) are expressible in radicals.

Using (a version of) the Galois solvability criterion 8.2.8 (a) below, one can prove that *for $a, b \in \mathbb{Q}$ all roots of the polynomial $x^5 + ax + b$ are expressible in radicals if and only if either the polynomial is reducible or*
$$a = \frac{15 \pm 20c}{c^2 + 1} \quad \text{and} \quad b = \frac{44 \mp 8c}{c^2 + 1} \quad \text{for some } c \in \mathbb{Q}, \ c \geq 0;$$
see [**PS97**, Ch. 6, section 7, Theorem 1].

8.2.8. (a)* **Galois solvability criterion (conjecture).**

For every $a_{n-1}, \ldots, a_0 \in \mathbb{Q}$, all the roots of the equation $A(x) := x^n + a_{n-1}x^{n-1} + \ldots + a_1 x + a_0 = 0$ are expressible in radicals if and only if one can obtain, starting from $\{A\}$, a set of polynomials of degree one over $\mathbb{Q}$ by using the following operations:

• (factorization) if one of our polynomials equals $P_1 P_2$ for some non-constant $P_1, P_2 \in \mathbb{Q}[x]$, then replace $P_1 P_2$ by P_1 and P_2;

• (extracting a root) if one of our polynomials equals $P(x^q)$ for some $P \in \mathbb{Q}[x]$, then replace $P(x^q)$ by $P(x)$;

• (taking the Galois resolvent) replace one of our polynomials P by the polynomial $Q(y_1, \ldots, y_q, y)$ where $y_1, \ldots, y_q$ are all the roots of P. (By problem 8.2.7 (c), $Q(y_1, \ldots, y_q, y) \in \mathbb{Q}[y]$.)

(b) Prove the "if" part of criterion (a).

(c)* State and prove the real analogue of criterion (a).

(d)* State and prove the analogue of criterion (a) for equations that are solvable using one radical; cf. [**Akh, ABG+**].

(e)* Does the analogue of (a) hold for every $a_{n-1}, \ldots, a_0 \in \mathbb{C}$ with "expressible in radicals" replaced by "expressible in radicals from $\{1, a_{n-1}, \ldots, a_0\}$"?

The proof of the "only if" part of criterion (a) is presumably similar to Theorem 8.1.13; see also the Galois Theorem 8.1.12 and subsection 3.I. I would be grateful if an expert in algebra could confirm that criterion (a) is correct (and is equivalent to the Galois solvability criterion in its standard textbook formulation), or describe the required changes. (I asked experts in July 2017, but so far have received no answer.)

Suggestions, solutions, and answers

8.2.3. (a) The polynomial $(x - y)^2 (y - z)^2 (z - x)^2$ is symmetric. (One may also reduce (a) to (b).)

(b) Set

$$M = x^9 y + y^9 z + z^9 x \quad \text{and} \quad N = y^9 x + x^9 z + z^9 y.$$

Then $M + N$ and MN are symmetric polynomials. Therefore they are polynomials in elementary symmetric polynomials $\sigma_1, \sigma_2, \sigma_3$. (An explicit expression is given in [**ABG+**].) Thus, M can be expressed once we know $M + N$ and MN; see the beginning of 2.A.

8.2.4. (a) One possible answer is $u + v\varepsilon_3 + w\varepsilon_3^2$.
   (b) Set

$$M = x_1 x_3 + x_3 x_5 + x_5 x_7 + x_7 x_9 + x_9 x_1 \quad \text{and}$$

$$N = x_2 x_4 + x_4 x_6 + x_6 x_8 + x_8 x_{10} + x_{10} x_2,$$

and proceed similarly to problem 8.2.3 (b).

8.2.5. (a) The square $(x - y)^2 (x - z)^2 (x - t)^2 (y - z)^2 (y - t)^2 (z - t)^2$ is symmetric; cf. problem 8.2.3 (a).
   (b) Set

$$M = xy + zt, \quad N = xz + yt, \quad K = xt + yz.$$

By 8.2.3 (c), M can be expressed in radicals using the polynomials

$$M + N + K, \quad MN + MK + NK, \quad MNK.$$

Proceeding similarly to problems 8.2.3 (c) and 8.2.5 (d), we see that these polynomials are symmetric. Thus $M = xy + zt$ is expressible in radicals.
   (c) Set

$$M = (x + y - z - t)^2, \quad N = (x + z - y - t)^2, \quad K = (x + t - y - z)^2$$

and repeat the solution of (b) to obtain $M = (x + y - z - t)^2$. Then it is easy to obtain $x + y - z - t$.
   Alternative solution. We have

$$(x + y - z - t)^2 = (x^2 + y^2 + z^2 + t^2) + 2(xy + tz) - 2(xt + yz) - 2(xz + yt).$$

The first summand is symmetric and the other summands are expressible in radicals by (b). Thus $x + y - z - t$ is expressible in radicals.

8.2.6. Repeatedly use the identities $1 + \varepsilon + \varepsilon^2 = 0$ and $1 + i + i^2 + i^3 = 0$.

8.2.7. Here we show the solution for $q = 5$.
   (a) We have

$$t(\varepsilon_5 \vec{u}_\alpha) = t(u_{\alpha(5)}, u_{\alpha(1)}, u_{\alpha(2)}, u_{\alpha(3)}, u_{\alpha(4)}) = t(\vec{u}_{\alpha \circ (54321)}).$$

Hence

$$Q(\varepsilon_5 u_1, \ldots, \varepsilon_5 u_5, y) = \prod_{\alpha \in \Sigma_5} (y - t(\varepsilon_5 \vec{u}_\alpha))$$

$$= \prod_{\alpha \in \Sigma_5} (y - t(\vec{u}_{\alpha \circ (54321)})) = Q(u_1, \ldots, u_5, y).$$

Here

- $(54321) \in \Sigma_5$ is the cycle that sends 5 to 4, 4 to 3, ..., 1 to 5;
- the last equality holds because when α ranges through Σ_5, so too does $\alpha \circ (54321)$.

(b) There exists a homogeneous polynomial $P_k \in \mathbb{Q}[\varepsilon_5][u_1, \ldots, u_5]$ (of "degree" $120 - k$) such that the coefficient of y^k in Q is $P_k(u_1, \ldots, u_5)$, i.e.,

$$Q(u_1, \ldots, u_5, y) = \sum_{k=0}^{120} P_k(u_1, \ldots, u_5) y^k.$$

By (a) and homogeneity we have

$$P_k(u_1, \ldots, u_5) = P_k(\varepsilon_5 u_1, \ldots \varepsilon_5 u_5) = \varepsilon_5^{-k} P_k(u_1, \ldots, u_5).$$

If k is not divisible by 5, we obtain $P_k(u_1, \ldots, u_5) = 0$ as required.

(c) The polynomial $Q(u_1, \ldots, u_5, y)$ is symmetric in $u_1, \ldots, u_5$. Thus all the coefficients (P_k in (b)) of the corresponding polynomial from $Q[\varepsilon_5, u_1, \ldots, u_5][y]$ are symmetric in $u_1, \ldots, u_5$. Now $Q(x_1, \ldots, x_5, y) \in \mathbb{Q}[\varepsilon_5][y]$ by the Fundamental Theorem on Symmetric Polynomials 3.6.3(d), Vieta's Theorem 3.6.5, and the fact that all coefficients of f are rational.

The assertion that $Q(x_1, \ldots, x_5, y) \in \mathbb{Q}[y]$ is proved similarly to the rationality lemmas 8.3.18 (f), 8.3.22 (d), and 8.4.17.

8.2.8. (b) Use Lagrange resolvents.

2.C. A reformulation of the constructibility in Gauss's Theorem (2)

A complex number is called *complex constructible* if it can be obtained, starting with 1, by finitely many operations of addition, subtraction, multiplication, division by a nonzero number, and taking square roots. More precisely, a complex number z is called constructible if some set of complex numbers containing z can be obtained from the one-element set $\{1\}$ using the following operations:

To a given set M and $x, y \in M$ one can add

- $x + y$, xy, and x/y for $y \neq 0$;
- any number $r \in \mathbb{C}$ such that $r^2 = x$.

8.2.9. The number $\cos \frac{2\pi}{n}$ is real constructible if and only if the number $\varepsilon_n := \cos(2\pi/n) + i\sin(2\pi/n)$ is complex constructible.

Lemma 8.2.10 (Complexification). A complex number is complex constructible if and only if its real and imaginary parts are real constructible.

It follows from this lemma that a real number is complex constructible if and only if it is real constructible.[4] Therefore, it suffices to prove Gauss's theorem 8.1.5 with "real constructibility" replaced by "complex constructibility".

8.2.11.* Is the number e $\{e + \pi i\}$-expressible in radicals? (Recall the definition preceding Theorem 8.1.13. You may use without proof the fact that e and π are not expressible in radicals.)

Suggestions, solutions, and answers

8.2.9. The statement follows from

$$\varepsilon_n = \cos\frac{2\pi}{n} + i\sin\frac{2\pi}{n} = \cos\frac{2\pi}{n} + \sqrt{-\sin^2\frac{2\pi}{n}} = \cos\frac{2\pi}{n} + \sqrt{\cos^2\frac{2\pi}{n} - 1},$$

$$\cos\frac{2\pi}{n} = \frac{\varepsilon_n + \varepsilon_n^{-1}}{2}, \quad \text{and} \quad \sin\frac{2\pi}{n} = \frac{\varepsilon_n - \varepsilon_n^{-1}}{2}$$

or from Lemma 8.2.10.

8.2.10. The "if" part is clear. In order to prove the "only if" part, write $\sqrt{a+bi} = u + vi$ and express u and v in terms of a and b using the four arithmetic operations and the square root operation.

2.D. Idea of the proof of constructibility in Gauss's Theorem (2)

8.2.12. (a) The number ε_5 is constructible.

(b) The number ε_7 can be obtained, starting from 1, by finitely many operations of addition, subtraction, multiplication, division by a nonzero number, and taking complex square and cube roots of complex numbers.

(c)* Prove (b) under the additional restriction that each type of root operation occurs only once (one instance of taking a square root and one instance of taking a cube root).

(d) The number ε_{11} is expressible in radicals including only square and fifth roots.

(e) The number ε_{17} is constructible.

---

[4]Note that our definition of constructibility does not include the functions Re and Im. It is possible to "realize" them by showing that if you can obtain the complex number z, then you can obtain $\bar{z}$. However, this will only prove the *complex constructibility* of the real and imaginary parts, but not their *real constructibility*. To prove the real constructibilty you need to extract the complex root using real roots. This is only possible for square roots. If in the definition of constructibility and real constructibility we were to allow the extraction of cube roots only, then the analogue of the complexification lemma would be incorrect. Indeed, $\varepsilon_9 \in \{\sqrt[3]{\sqrt[3]{1}}\}$ is expressible in radicals if we allow cube roots, but $\cos\frac{2\pi}{9}$ is not expressible in real radicals; see Remark 8.1.7 (f).

Parts (a), (b), and (c) can be solved directly (for clues to (b, c) see the hints for problem 8.3.14 (b)). For (d) and (e) you will need a new idea, outlined in 2.B and below. Instead of working with a set of roots, it is more convenient to work with the *Lagrange resolvents* defined in 2.B.

Sketch of the proof of the constructibility of $\varepsilon := \varepsilon_5$**.** First note that

$$T_0 := \varepsilon + \varepsilon^2 + \varepsilon^4 + \varepsilon^8 = -1.$$

We begin by proving the constructibility of

$$T_2 := \varepsilon - \varepsilon^2 + \varepsilon^4 - \varepsilon^8.$$

If we substitute ε^2 for ε, then T_2 goes to $-T_2$. Thus, T_2^2 is invariant under this substitution. Therefore T_2^2 will not change after repeating this substitution—in other words, after replacing ε with ε^4 or with $\varepsilon^8 = \varepsilon^3$. So, for any k, the number T_2^2 will not change if we replace ε with ε^k.

Expand the brackets in the product $T_2^2 = T_2 \cdot T_2$ and replace ε^5 with 1. We get the equality

$$T_2^2 = a_0 + a_1\varepsilon + a_2\varepsilon^2 + a_3\varepsilon^3 + a_4\varepsilon^4 \quad \text{for some } a_k \in \mathbb{Z}.$$

Since T_2^2 does not change when ε is replaced by ε^k, we have $a_1 = a_2 = a_3 = a_4$. Therefore $T_2^2 = a_0 - a_1 \in \mathbb{Z}$, implying that T_2 is constructible.

Let

$$T_1 := \varepsilon + i\varepsilon^2 - \varepsilon^4 - i\varepsilon^8 \quad \text{and} \quad T_3 := \varepsilon - i\varepsilon^2 - \varepsilon^4 + i\varepsilon^8.$$

Then $T_0 + T_1 + T_2 + T_3 = 4\varepsilon$. Thus it suffices to prove that T_1 and T_3 are constructible. We will prove this for T_1; the proof for T_3 is similar. If we replace ε by ε^2, then T_1 goes to $-iT_1$. Thus, T_1^4 does not change under this substitution. Similarly, T_1^4 will not change after repeating the substitution, i.e., after replacing ε with ε^4 or $\varepsilon^8 = \varepsilon^3$. So, for any k, the number T_1^4 does not change when ε is replaced with ε^k.

As above, we get

$$T_1^4 = a_0 + a_1\varepsilon + a_2\varepsilon^2 + a_3\varepsilon^3 + a_4\varepsilon^4 \quad \text{for some } a_k \in \mathbb{Z} + i\mathbb{Z}.$$

Since T_1^4 does not change when ε is replaced by ε^k, we see that $a_1 = a_2 = a_3 = a_4$. Therefore $T_1^4 = a_0 - a_1 \in \mathbb{Z} + i\mathbb{Z}$. It follows that T_1 is constructible. $\qquad \square$

In the above arguments, we had to conclude that $a_1 = a_2 = a_3 = a_4$ and define carefully what "replacing ε with ε^2" means. The proof for the general case is difficult; the reader can find an example of such arguments in [**Edw97**, Ch. 24]. Instead, we slightly modify our proof; rather than working with numbers, we will work with polynomials and take their values at ε. Two polynomials with complex coefficients are said to be *congruent modulo the polynomial p* if their difference is divisible by p (in $\mathbb{C}[x]$).

8.2.13. Let $T_1(x) := x + ix^2 - x^4 - ix^8$. Then
    (a) $iT_1(x^2) \equiv T_1(x) \bmod (x^5 - 1)$;

(b) $T_1^4(x^2) \equiv T_1^4(x) \bmod (x^5 - 1)$;

(c) $T_1^4(x^k) \equiv T_1^4(x) \bmod (x^5 - 1)$ for any k.

A proof of constructibility of $\varepsilon := \varepsilon_5$. Define the polynomial $T_1(x) := x + ix^2 - x^4 - ix^8$. Define polynomials $T_0(x)$, $T_2(x)$, and $T_3(x)$ similarly following the definitions of the numbers T_0, T_2, and T_3 in the arguments on the previous page. As above, $(T_0 + T_1 + T_2 + T_3)(\varepsilon) = 4\varepsilon$. Therefore it suffices to prove the constructibility of each of the numbers $T_r(\varepsilon)$ for $r = 1, 2, 3$. We have

$$iT_1(x^2) \underset{x^5-1}{\equiv} T_1(x) \implies T_1^4(x^2) \underset{x^5-1}{\equiv} T_1^4(x)$$

$$\implies T_1^4(x^k) \underset{x^5-1}{\equiv} T_1^4(x) \quad \text{for any } k.$$

Consider the polynomial $a_0 + a_1 x + a_2 x^2 + a_3 x^3 + a_4 x^4$ with coefficients in $\mathbb{Z} + i\mathbb{Z}$ congruent to $T_1^4(x)$ modulo $x^5 - 1$.

Then $a_1 = a_2 = a_3 = a_4$. Therefore $T_1^4(\varepsilon) = a_0 - a_1 \in \mathbb{Z} + i\mathbb{Z}$. Thus $T_1(\varepsilon)$ is constructible.[5] Similarly $T_2(\varepsilon)$ and $T_3(\varepsilon)$ are also constructible. $\square$

8.2.14. (a) Let

$$\beta := \varepsilon_6 = \frac{1 + i\sqrt{3}}{2} \quad \text{and} \quad T(x) := x + \beta x^3 + \beta^2 x^9 + \beta^3 x^{27} + \beta^4 x^{81} + \beta^5 x^{243}.$$

Prove that $T(x) \equiv \beta T(x^3) \bmod (x^7 - 1)$.

(b) Let

$$\beta := \varepsilon_{10} \quad \text{and} \quad T(x) := x + \beta x^2 + \beta^2 x^4 + \beta^3 x^8 + \beta^4 x^{16} + \ldots + \beta^9 x^{512}.$$

Prove that $T(x) \equiv \beta T(x^2) \bmod (x^{11} - 1)$.

The solutions of problems 8.2.12 (d, e) and 8.2.14 are similar to the proof of the constructibility of ε_5. For details see 2.E.

2.E. Proof of the constructibility in Gauss's Theorem (3)

Note that formally the proof we gave is independent of 2.D, and from 2.C we only used the complexification lemma 8.2.10.

Lemma 8.2.15 (Multiplication). (a) If ε_n is constructible, then ε_{2n} is constructible.

(b) If ε_m and ε_n are constructible and m and n are relatively prime, then ε_{mn} is constructible.

---

[5] Here is another proof suggested by M. Yagudin:

$$T_1^4(\varepsilon) = a_0 + a_1\varepsilon + a_2\varepsilon^2 + a_3\varepsilon^3 + a_4\varepsilon^4 = a_0 + a_1\varepsilon^2 + a_2\varepsilon^4 + a_3\varepsilon + a_4\varepsilon^3$$

$$= a_0 + a_1\varepsilon^3 + a_2\varepsilon + a_3\varepsilon^4 + a_4\varepsilon^2 = a_0 + a_1\varepsilon^4 + a_2\varepsilon^3 + a_3\varepsilon^2 + a_4\varepsilon.$$

Summing these expressions, we get $4T_1^4(\varepsilon) = a_0 - a_1 - a_2 - a_3 - a_4 \in \mathbb{Z} + i\mathbb{Z}$.

Proof. This follows from the formulas $\varepsilon_{2n} \in \{\sqrt{\varepsilon_n}\}$ and $\varepsilon_{mn} = \varepsilon_m^x \varepsilon_n^y$, where x and y are integers such that $mx + ny = 1$. □

In problem 8.2.12 (a), we used the difference in the remainders upon dividing the numbers 2, 2^2, 2^3, and 2^4 by 5. In problems 8.2.12 (d, e) and 8.2.14 (a) we used similar properties of the numbers 2 and 11, 6 and 17, and 3 and 7. For the general case, the following generalization is needed.

Theorem 8.2.16 (Primitive roots). For every prime p there exists an integer g such that the residues modulo p of $g^1, g^2, g^3, \ldots, g^{p-1}$ are all distinct.

Sketch of proof for the case $p = 2^m + 1$ (the only case used by Gauss's Theorem). If there are no primitive roots, then the congruence $x^{2^{m-1}} \equiv 1 \bmod p$ has $p - 1 = 2^m > 2^{m-1}$ solutions. This contradicts Bezout's Theorem. For a complete proof, see section 5 in Chapter 2.

Proof of constructibility in Gauss's Theorem 8.1.5. By the complexification and multiplication lemmas (Lemmas 8.2.10 and 8.2.15), it suffices to prove that ε_n is constructible for any prime $n = 2^{2^s} + 1$. Since $n - 1 = 2^m$, the multiplication lemma 8.2.15 shows that $\beta := \varepsilon_{n-1}$ is constructible. Define

$$\mathbb{Z}[\beta] := \{a_0 + a_1\beta + a_2\beta^2 + \ldots + a_{n-2}\beta^{n-2} \mid a_0, \ldots, a_{n-2} \in \mathbb{Z}\}.$$

Let g be a primitive root modulo n. For $r = 0, 1, 2, \ldots, n - 2$, define

$$T_r(x) := x + \beta^r x^g + \beta^{2r} x^{g^2} + \ldots + \beta^{(n-2)r} x^{g^{n-2}} \in \mathbb{Z}[\beta][x].$$

Then $(T_0 + T_1 + \ldots + T_{n-2})(\varepsilon) = (n-1)\varepsilon$. Furthermore, $T_0(\varepsilon) = -1$. Therefore it suffices to prove the constructibility of each $T_r(\varepsilon), r = 1, 2, \ldots, n - 2$. We have

$$\beta^r T_r(x^g) \underset{x^n - 1}{\equiv} T_r(x) \Longrightarrow T_r^{n-1}(x^g) \underset{x^n - 1}{\equiv} T_r^{n-1}(x)$$

$$\Longrightarrow T_r^{n-1}(x^k) \underset{x^n - 1}{\equiv} T_r^{n-1}(x) \quad \text{for any } k.$$

Consider the polynomial $x + a_2 x^2 + \ldots + a_{n-1} x^{n-1}$ with coefficients in $\mathbb{Z}[\beta]$ that is congruent to $T_r^{n-1}(x)$ modulo $x^n - 1$. Then $a_1 = a_2 = \ldots = a_{n-1}$. Therefore $T_r^{n-1}(\varepsilon) = a_0 - a_1 \in \mathbb{Z}[\beta]$, which implies that $T_r(\varepsilon)$ is constructible. □

2.F. Efficient proofs of constructibility (4*)

Here are alternative proofs, due to Gauss, of constructibility and the degree-reducing theorem (8.1.15 (a)). They are more complicated than those given in 2.E, but provide faster computational algorithms (cf. [**BK13, Saf, Kog**]).

An "efficient" constructibility proof for $n = 5$. It suffices to prove that $\varepsilon := \varepsilon_5$ is constructible. Since it is difficult to immediately express ε in radicals, we shall first prove that *certain polynomials* of ε are constructible. We have $1 + \varepsilon + \varepsilon^2 + \varepsilon^3 + \varepsilon^4 = 0$. Hence

$$(\varepsilon + \varepsilon^4)(\varepsilon^2 + \varepsilon^3) = \varepsilon + \varepsilon^2 + \varepsilon^3 + \varepsilon^4 = -1.$$

Define

$$T_0 := \varepsilon + \varepsilon^4 \quad \text{and} \quad T_1 := \varepsilon^2 + \varepsilon^3.$$

Then T_0 and T_1 are roots of the equation $t^2 + t - 1 = 0$ by Vieta's Theorem 3.6.5. Hence these numbers are constructible. Likewise, since $\varepsilon \cdot \varepsilon^4 = 1$, the numbers ε and ε^4 are roots of the equation $t^2 - T_0 t + 1 = 0$, so ε (and ε^4) are constructible. $\square$

Sketch of the proof of the degree-lowering theorem 8.1.15 (a). (For $n-1 = 2^m$ we obtain an idea of the proof of constructibility in the Gauss Theorem 8.1.5.) Factor $n-1$ into primes $q_1 q_2 \cdots q_s$. First, it would be nice to partition the sum

$$\varepsilon_n + \varepsilon_n^2 + \ldots + \varepsilon_n^{n-1} = -1$$

into q_1 terms $T_0, T_1, \ldots, T_{q_1-1}$ that are expressible in radicals (in other words, cleverly group the roots of the equation $1 + x + x^2 + \ldots + x^{n-1} = 0$). Then we would partition each T_k into q_2 terms $T_{k,0}, T_{k,1}, \ldots, T_{k,q_2-1}$ that are expressible in radicals, etc., until we get $T_{\underbrace{1,\ldots,1}_{s}} = \varepsilon_n$.

However, finding these clever groupings of the numbers $1, \varepsilon_n, \varepsilon_n^2, \ldots, \varepsilon_n^{n-1}$ is not a trivial task.

8.2.17. Partition
    (a) $\varepsilon_7, \varepsilon_7^2, \ldots, \varepsilon_7^6$ into 2 groups of 3 elements;
    (b) $\varepsilon_{11}, \varepsilon_{11}^2, \ldots, \varepsilon_{11}^{10}$ into 2 groups of 5 elements;
    (c) $\varepsilon_{13}, \varepsilon_{13}^2, \ldots, \varepsilon_{13}^{12}$ into 2 groups of 6 elements
so that each group has a constructible sum.

The primitive root theorem 8.2.16 allows one to encode nonzero residues modulo the prime n using residues modulo $n-1$. Namely, choosing a primitive root g, we associate a residue k modulo $n-1$ with the nonzero remainder upon dividing g^k by n. This encoding was actually used in the groupings constructed above for $n = 5$ and in problem 8.2.17.

We now sketch an "efficient" proof of constructibility in Gauss's Theorem 8.1.5. It suffices to prove the constructibilty of $\varepsilon := \varepsilon_n$ for a prime $n = 2^m + 1 \geq 5$. Up until problem 8.2.21 we assume that n satisfies this condition. Let g be a primitive root modulo n. For $n = 5$ we partitioned into terms of the form ε^k, where the k's are even and odd powers of the primitive root g respectively (that is, quadratic residues and nonresidues modulo n). Generalizing, we define

$$T_0 := \varepsilon^{g^2} + \varepsilon^{g^4} + \varepsilon^{g^6} + \cdots + \varepsilon^{g^{2^m}} \quad \text{and} \quad T_1 := \varepsilon^{g^1} + \varepsilon^{g^3} + \varepsilon^{g^5} + \cdots + \varepsilon^{g^{2^m-1}}.$$

We now show that T_0 and T_1 are constructible. As before, $T_0 + T_1 = -1$. Therefore it suffices to check that the product $T_0 T_1$ is an integer.

8.2.18. Prove the following equalities.

(a) $T_0 T_1 = \sum_{s=0}^{2^m} \varepsilon^s N_s$, where N_s is the number of solutions (in residues modulo 2^{m-1}) of the congruence $g^{2k} + g^{2l+1} \equiv s \pmod{n}$, that is, the number of ways to represent a residue modulo n as the sum $u + v$, where ε^u is a term in the sum T_0 and ε^v is a term in the sum T_1:

$$N_s = \{(b, c) \in \mathbb{Z}_{2m}^2 \mid b \equiv 0,\ c \equiv 1 \bmod 2,\ g^b + g^c \equiv s \bmod n\};$$

(b) $N_0 = 0$;

(c) $N_s = N_{gs}$ when $s \neq 0$; i.e., the congruence $g^{2k} + g^{2l+1} \equiv s \bmod n$ has as many solutions (k, l) (in residues modulo 2^m) as does the congruence $g^{2k} + g^{2l+1} \equiv gs \bmod n$;

(d) $N_1 = N_2 = \ldots = N_{2m}$;

(e) $T_0 T_1 = -\frac{n-1}{4} = -2^{m-2}$.

8.2.19. Find the number of representations of the residue of 2017 modulo $p := 10^9 + 9$ as the sum of a quadratic residue modulo p and a quadratic nonresidue modulo p. (Your answer should be an integer, not a formula or an algorithm. You may use without proof the fact that p is prime.)

Now it is clear how to continue the proof; for the next groupings we will use congruences modulo 4, etc.

8.2.20. (a) The congruence $4k + g^{4l+2} \equiv 1 \bmod n$ has as many solutions (k, l) (in residues modulo 2^{m-2}) as does the congruence $g^{4k} + g^{4l+2} \equiv g^2 \bmod n$.

(b) Define

$$T_{00} := \varepsilon^{g^4} + \varepsilon^{g^8} + \varepsilon^{g^{12}} + \cdots + \varepsilon^{g^{2m}} \quad \text{and} \quad T_{01} := \varepsilon^{g^2} + \varepsilon^{g^6} + \varepsilon^{g^{10}} + \cdots + \varepsilon^{g^{2m-2}}.$$

Prove that $T_{00} T_{01} = s T_0 + t T_1$ for some integers s and t.

(c) Define

$$T_{11} := \varepsilon^{g^1} + \varepsilon^{g^5} + \varepsilon^{g^9} + \cdots + \varepsilon^{g^{2m-3}} \quad \text{and} \quad T_{10} := \varepsilon^{g^3} + \varepsilon^{g^7} + \varepsilon^{g^{11}} + \cdots + \varepsilon^{g^{2m-1}}.$$

Prove that $T_{10} T_{11} = u T_0 + v T_1$ for some integers u and v.

An "efficient" proof of constructibility in Gauss's Theorem 8.1.5. It suffices to prove the constructibility of $\varepsilon := \varepsilon_n$ for a prime $n = 2^m + 1$. Let $a_j \in \mathbb{Z}_2 = \{0, 1\}$ for each $j \in \{0, 1, 2, \ldots, m-1\}$. For each $j \in \{0, 1, 2, \ldots, m-1\}$ define

$$\overline{a_j \ldots a_1 a_0} := a_0 + 2a_1 + 2^2 a_2 + \ldots + a_j 2^j,$$

i.e., the binary representation. It is important to note that this binary representation can start with zeros.

Let g be a primitive root modulo n. For $A \in \mathbb{Z}_2^k$ let

$$T_A := \sum_{0 \leq b \leq n-1,\ b \equiv \overline{A} \bmod 2^k} \varepsilon^{g^b}.$$

Using induction on k, we prove that for any k and $A \in \mathbb{Z}_2^k$, the number T_A is constructible. Then for $k = m$ we obtain the constructibility of $T_{\underbrace{0 \ldots 0}_{m}} = \varepsilon$.

The base case $k = 0$ follows from $T_\varnothing = -1$, where the subscript denotes not the empty set but the vector of length zero. For the inductive step, suppose that for some k the statement is true. Take any $A \in \mathbb{Z}_2^k$. Then $T_A = T_{0A} + T_{1A}$.[6] Furthermore,

$$T_{0A}T_{1A} = \sum_{s=0}^{n-1} N_s \varepsilon^s \overset{(*)}{=} N_0 + \sum_{c=0}^{2^j-1} \sum_{\substack{0 \leq b \leq n-1, \\ b \equiv c \bmod 2^j}} N_{g^b} \varepsilon^{g^b} = N_0 + \sum_{C \in \mathbb{Z}_2^k} N_{g^{\overline{C}}} T_C.$$

Here N_s (which depends on A) is the number of ways to express the residue s modulo n as the sum $k + l$, where ε^k is a term in the sum of T_{0A} and ε^l is a term in the sum of T_{1A}:

$$N_s = \left\{ (b_0, b_1) \in \mathbb{Z}_{n-1}^2 \,\middle|\, b_0 \underset{2^{k+1}}{\equiv} \overline{A}, \ b_1 \underset{2^{k+1}}{\equiv} 2^k + \overline{A}, \ g^{b_0} + g^{b_1} \underset{n}{\equiv} s \right\}.$$

The congruence $g b_0 + g^{b_1} \underset{n}{\equiv} s$ can be multiplied by g^{2^j} (similarly to the case involving T_0 and T_1 above), yielding $g^{b'_1} + g b'_0 = d^{2^j} s$ where

$$b'_0 = b_1 + 2^k \underset{2^{k+1}}{\equiv} \overline{A}, \qquad b'_1 = b_0 + 2^j \underset{2^{k+1}}{\equiv} 2^k + \overline{A}.$$

Since we can multiply by inverses as well, $N_s = N_{g^{2^k} s}$, which implies the equality $(*)$. Therefore T_{0A} and T_{1A} are constructible.

---

[6]The end of this paragraph can be replaced by the following slightly more complicated reasoning, which is better suited to obtaining a generalization of Gauss's degree-lowering theorem 8.1.15 (a). Instead of $T_{0A}T_{1A}$, consider

$$(T_{0A} - T_{1A})^2 = \left(\sum_{Bl \in \mathbb{Z}_2^{m-k}} (-1)^l \varepsilon^{g^{\overline{BlA}}} \right)^2 = \sum_{l=0,s=0}^{1,n-1} N_{s,l} (-1)^l \varepsilon^s$$

$$\overset{(**)}{=} \sum_{l=0}^{1} (-1)^l \left(N_{0,l} + \sum_{C \in \mathbb{Z}_2^k} N_{g^{\overline{C}},l} T_C \right).$$

Here $N_{s,l}$ (depending on A) is the number of ordered solutions $l_1, l_2 \in \mathbb{Z}_2$, $B_1, B_2 \in \mathbb{Z}_2^{m-k-1}$ to the system of congruences

$$\begin{cases} l_1 + l_2 \equiv l \bmod 2, \\ g^{\overline{B_1 l_1 A}} + g^{\overline{B_2 l_2 A}} \equiv s \bmod n. \end{cases}$$

Clearly, $N_{s,l} = N_{g^{2^k} s, l}$, which implies equality $(**)$.

8.2.21. Find explicit formulas for the following numbers, using only square, cube, and fifth roots:
  (a) ε_7;   (b) $\varepsilon_{13} + \varepsilon_{13}^3 + \varepsilon_{13}^9$;   (c) ε_{13};   (d) ε_{11}.

An "efficient" constructibility proof of the degree-lowering theorem 8.1.15 (a). Let $q_1 q_2 \cdots q_m$ be the prime factorization of $n-1$ into not necessarily distinct primes. Let

$$a_i \in \{0, 1, 2, \ldots, q_{i+1} - 1\} \quad \text{for each } i \in \{0, 1, 2, \ldots, m-1\}.$$

Define

$$\overline{a_{m-1} \cdots a_1 a_0} := a_0 + a_1 q_1 + a_2 q_1 q_2 + \ldots + a_{m-1} q_1 q_2 \cdots q_{m-1},$$

i.e., the "variable base" representation. As above, note that this may start with zero "digits." Define

$$[k, l] := \mathbb{Z}_{q_k} \times \ldots \times \mathbb{Z}_{q_l}$$

to be a set of arrays of $k - l + 1$ "digits" that can be in the representation $\overline{a_{m-1} \cdots a_1 a_0}$ from the kth "digit" from the right to the lth "digit" from the right (where a_0 is considered to be the first digit from the right). Let g be a primitive root modulo n, and let $\varepsilon = \varepsilon_n$. For $A \in [k, l]$ we write

$$T_A := \sum_{B \in [m, k+1]} \varepsilon^{g^{\overline{BA}}}.$$

By induction on k we will demonstrate how to express T_A in radicals for any k and $A \in [l, k]$. Thus, for $k = m$ we will obtain an expression for $T_{\underbrace{0 \ldots 0}_{m}} = \varepsilon$ in radicals.

The base case $k = 0$ follows from $T_\emptyset = -1$. For the inductive step, let $q := q_{k+1}$ and $\beta := \varepsilon_q$. For any $r = 0, 1, 2, \ldots, q - 1$ and $A \in [k, l]$, define

$$T_A^{(r)} := T_{0A} + \beta^r T_{1A} + \beta^{2r} T_{2A} + \ldots + \beta^{(q-1)r} T_{(q-1)A}.$$

Then

$$T_A^{(0)} = T_A \quad \text{and} \quad q T_{lA} = \beta^{-l} T_A^{(0)} + \beta^{-2l} T_A^{(1)} + \ldots + \beta^{-(q-1)l} T_A^{(q-1)}.$$

For any $r = 0, 1, 2, \ldots, q - 1$ and $A \in [k, l]$, we have

$$(T_A^{(r)})^q = \left(\sum_{Bl \in [m, k+1]} \beta^{lr} \varepsilon^{g^{\overline{BlA}}} \right)^q = \sum_{l=0, s=0}^{q-1, n-1} N_{|s, l|} \varepsilon^s \beta^l$$

$$\overset{(***)}{=} \sum_{l=0}^{q-1} \beta^l \left(N_{|0, l|} + \sum_{C \in [k, 1]} N_{|g^{\overline{C}}, l|} T_C \right).$$

Here $N_{s,l}$ (which depends on A) is the number of ordered solutions $l_1, l_2, \ldots,$ $l_q \in \mathbb{Z}_q$, $B_1, B_2, \ldots, B_q \in [m, k+2]$ to the system of congruences

$$\begin{cases} r(l_1 + l_2 + \ldots + l_q) \equiv l \mod q, \\ g^{\overline{B_1 l_1 A}} + g^{\overline{B_2 l_2 A}} + \ldots + g^{\overline{B_q l_q A}} \equiv s \mod n. \end{cases}$$

Clearly, $N_{s,l} = N_{g^{q_1 q_2 \cdots q_k} s, l}$, implying ($*$). Thus T_{lA} is expressible in radicals.

Suggestions, solutions, and answers

8.2.17. (a) Let $\varepsilon := \varepsilon_7$. Obtain the expressions

$$T_0 := \varepsilon^{3^0} + \varepsilon^{3^2} + \varepsilon^{3^4} \quad \text{and} \quad T_1 := \varepsilon^3 + \varepsilon^{3^3} + \varepsilon^{3^5}.$$

An "inefficient" proof of expressibility in radicals. The quantity $T_0 T_1$ is a polynomial in ε with integer coefficients of degree less than 7 (more precisely, it is the value at $x = \varepsilon$ of some polynomial in x (modulo $x^7 - 1$) with integer coefficients). The substitution $\varepsilon \to \varepsilon^3$ interchanges T_0 and T_1 and leaves $T_0 T_1$ unchanged. Hence the coefficient of a polynomial at ε^s is equal to its coefficient at ε^{3s}. Since 3 is a primitive root modulo 7, all the coefficients of the polynomial, except for the constant term, are equal. From this and from $\varepsilon + \varepsilon^2 + \ldots + \varepsilon^6 = -1$ it follows that $T_0 T_1$ is an integer. Therefore T_0 and T_1 are expressible in radicals.

An "efficient" proof of expressibility in radicals. We have

$$T_0 T_1 = \sum_{s=0}^{6} N_s \varepsilon^s,$$

where N_s is the number of solutions $(n, m) \in \mathbb{Z}_3^2$ of the congruence $3^{2n} + 3^{2m+1} \equiv s \mod 7$. It is clear that $N_0 + N_1 + N_2 + \ldots + N_6 = 9$. It is easy to verify that $N_s = N_{3s}$. Therefore $N_1 = N_2 = \ldots = N_6$. Since $3^0 + 3^1 \not\equiv 0 \mod 7$, we have $N_0 \neq 9$. It follows that $N_0 = 3$ and $N_1 = 1$, so $T_0 T_1 = 3 - 1 = 2$. Thus $\{T_0, T_1\} = \left\{ \frac{-1-\sqrt{7}i}{2}, \frac{-1+\sqrt{7}i}{2} \right\}$.

(c) Define $\varepsilon := \varepsilon_{13}$,

$$T_0 := \varepsilon + \varepsilon^3 + \varepsilon^4 + \varepsilon^{12} + \varepsilon^9 + \varepsilon^{10}, \quad \text{and} \quad T_1 := \varepsilon^2 + \varepsilon^5 + \varepsilon^6 + \varepsilon^7 + \varepsilon^8 + \varepsilon^{11}.$$

Then $T_0 + T_1 = -1$ and $T_0 T_1 = -3$. Thus

$$\{T_0, T_1\} = \left\{ \frac{-1 - \sqrt{13}}{2}, \frac{-1 + \sqrt{13}}{2} \right\}.$$

Estimating the value of T (for example, by a careful examination or by looking at the regular 13-gon), we get $T_1 + 1 < 0$. Thus $T_1 = \frac{-1-\sqrt{13}}{2}$ and $T_0 = \frac{-1+\sqrt{13}}{2}$.

8.2.18. (b) We have $g^k + g^l \equiv 0 \bmod n$ if and only if $k - l \equiv 2^{m-1} \bmod 2^m$.

(c) Multiplying any representation $g^b + g^c \equiv s \bmod n$ by g, we get a representation $g^{c'} + g^{b'} \equiv gs \bmod n$ where $b' = c + 1 \equiv 0 \bmod 2$ and $c' = b + 1 \equiv 1 \bmod 2$. Thereby $N_{gs} = N_s$ for all s.

(d) Since g is a primitive root, it follows from (c) that all N_s for $s \neq 0$ are equal.

(e) In view of (a, b, c, d) we have $T_0 T_1 = N_0 - N_1 \in \mathbb{Z}$.

8.2.21. (a) Let $\varepsilon := \varepsilon_7$. Having calculated T_0 and T_1 in problem 8.2.17 above, we can obtain ε. Define

$$\beta := \varepsilon_3 \quad \text{and} \quad T_{01} := \varepsilon^{3^0} + \beta \varepsilon^{3^2} + \beta^2 \varepsilon^{3^4}.$$

An "inefficient" proof of expressibility. Note that T_{01}^3 is a polynomial in ε with coefficients in $\mathbb{Z}[\beta]$ of degree less than 7. The substitution $\varepsilon \to \varepsilon^{3^2}$ leaves T_{01}^3 unchanged. Thus the coefficient ε^s of this polynomial is equal to the coefficient at $\varepsilon^{3^2 s}$. Since 3 is a primitive root modulo 7, coefficients of the polynomial at powers of 3^{2n} are equal and coefficients of the polynomial at powers of 3^{2n+1} are also equal. It follows that T_{01}^3 can be expressed using cube roots of numbers from $\mathbb{Z}[\beta, T_0, T_1]$. Thus T_{01} is expressible.

An "efficient" proof of expressibility. We have

$$T_{01}^3 = \sum_{s=1}^{7} \sum_{l=0}^{2} N_{s,l} \varepsilon^s \beta^l,$$

where $N_{s,l}$ is the number of solutions $(l_1, l_2, l_3) \in \mathbb{Z}_3^3$ of the system of congruences

$$\begin{cases} l_1 + l_2 + l_3 \equiv l \bmod 3, \\ 3^{2l_1} + 3^{2l_2} + 3^{2l_3} \equiv s \bmod 7. \end{cases}$$

It is clear that $\sum_{s=1}^{7} \sum_{l=0}^{2} N_{s,l} = 27$. It is easy to check that $N_{s,l} = N_{3^2 s,l}$. Therefore $N_{1,l} = N_{2,l} = N_{4,l}$ and $N_{3,l} = N_{5,l} = N_{6,l}$. Thus $T_{01}^3 = u + vT_0 + wT_1$ for some $u, v, w \in \mathbb{Z}[\beta]$. These are easy to find.

Last hint. Similarly, $T_{02} := \varepsilon^{3^1} + \beta \varepsilon^{3^3} + \beta^2 \varepsilon^{3^5}$ is expressible. Thus $\varepsilon = \frac{T_0 + T_{01} + T_{02}}{3}$ is expressible.

3. Problems on insolvabilty in radicals

In this section we use simple examples to illustrate the ideas of the proofs of the theorems on insolvability from section 1. This section is independent of the previous one. Moreover, it is almost independent of section 1 since most of the problems presented here concern the non-representability of numbers in various forms and do not use the definitions and formulations from section 1. Non-representability, although very natural, is not trivial to prove!

The problems in subsections 3.A–3.B lead to Theorem 8.1.2 and to the non-constructibility in Gauss's Theorem 8.1.5 (subsection 4.D). The problems from 3.A–3.D lead to the insolvability in real radicals from Theorem 8.1.8 (subsection 4.E), to Theorem 8.1.10 (subsection 4.H), and to Kronecker's Theorem 8.1.14 (subsection 4.G). The problems in 3.A–3.G lead to Theorem 8.1.13 and the insolvability in criteria 8.2.8 (a) (subsections 4.C and 4.F). The problems of this entire section, including subsection 3.I, lead to Theorem 8.1.16.

Subsections 3.A–3.D develop the idea of conjugation, and subsections 3.E–3.I develop the idea of symmetry. Note that in section 3 these ideas are revealed in the reverse order (since, in contrast to the first steps, the final realization of the idea of conjugation is more complicated than the idea of symmetry).

In this section, the term "polynomial" is shorthand for "polynomial with rational coefficients." Complex numbers $v_1, \ldots, v_n \in \mathbb{C}$ are called *linearly dependent* over $\mathbb{Q}$ if there exist numbers $\lambda_1, \ldots, \lambda_n \in \mathbb{Q}$, not all equal to zero, for which $\lambda_1 v_1 + \ldots + \lambda_n v_n = 0$. Recall that

$$\varepsilon_q := \cos(2\pi/q) + i\sin(2\pi/q).$$

3.A. Representability using only one square root (1–2)

Before attempting to solve the problems of this subsection, it is useful to work through the problems in section 1.

8.3.1. Can the following numbers be represented as $a + \sqrt{b}$ with $a, b \in \mathbb{Q}$?
  (a) $\sqrt{3 + 2\sqrt{2}}$;   (b) $\frac{1}{7+5\sqrt{2}}$;   (c) $\sqrt[3]{7 + 5\sqrt{2}}$;   (d) $\sqrt[3]{2}$;
  (e) $\sqrt{2} + \sqrt[3]{2}$;   (f) $\sqrt{2 + \sqrt{2}}$;   (g) $\sqrt{2} + \sqrt{3} + \sqrt{5}$.

Problems 8.3.1 and 8.3.3 are interesting in connection with insolvability in radicals because we need to come up with a polynomial whose roots are not radicals, and the numbers from problems 8.3.1 are the roots of polynomials (which ones?). See also 6.6.2.

Lemma 8.3.2 (Extension). Let α be a number obtained, starting with 1, by finitely many operations of addition, subtraction, multiplication, and division by a nonzero number and exactly one operation of taking the square root of a positive number. Then α is of the form $\alpha = a \pm \sqrt{b}$ for some $a, b \in \mathbb{Q}$ with $b > 0$.

8.3.3.* For which n can $\cos(2\pi/n)$ be represented in the form $a + \sqrt{b}$ where $a, b \in \mathbb{Q}$?
  Start with the cases $n = 16, 24, 20, 15, 9, 7, 17, 25$.

(The answer to this question is needed in the study of *outer billiards*. Compare with problem 3.1.7 (a), Remark 8.1.1, statement 8.1.3, and Theorems 3.8.5 and 8.1.5.)

Lemma 8.3.4. Assume that $r \in \mathbb{R} - \mathbb{Q}$ and $r^2 \in \mathbb{Q}$.

(a) **Irreducibility.** The polynomial $x^2 - r^2$ is irreducible over $\mathbb{Q}$.

(b) **Linear independence.** If $a, b \in \mathbb{Q}$ and $a + br = 0$, then $a = b = 0$.

(c) If r is a root of a polynomial, then this polynomial is divisible by $x^2 - r^2$.

(d) **Conjugation.** If r is a root of a polynomial, then $-r$ is also a root of this polynomial.

(e) **Conjugation.** If $a, b \in \mathbb{Q}$ and a polynomial has a root $a + br$, then $a - br$ is also a root of this polynomial.

(f) If $a, b \in \mathbb{Q}$ and a cubic polynomial has a root $a + br$, then this polynomial has a rational root.

Theorem 8.3.5. If a polynomial of degree at least 3 is irreducible over $\mathbb{Q}$, then none of its roots has the form $a \pm \sqrt{b}$ with $a, b \in \mathbb{Q}$.

Theorem 8.3.5 and Lemma 8.3.2 imply that *if a polynomial of degree at least 3 is irreducible over $\mathbb{Q}$, then none of its roots is expressible in real radicals by extracting just one square root.* The complex analogue of this statement is also true. This is our first step towards the theorems on insolvability in section 1. Similar general statements for higher-degree polynomials will be obtained below (formulate them yourself).

8.3.6. (a) Solve $x^6 - 2x^4 - 12x^3 - 2x^2 + 1 = 0$.

(b) The number $\cos(2\pi/7)$ is a root of the polynomial obtained from the function $x^3 + x^2 + x + 1 + x^{-1} + x^{-2} + x^{-3}$ by Zhukovsky's substitution $z = \frac{1}{2}(x + \frac{1}{x})$.

8.3.7. Is the polynomial $x^5 - 4x^3 + 6x^2 + 4x + 2$ reducible

(a) over $\mathbb{Z}$?   (b) over $\mathbb{Q}$?

8.3.8. (a) For each $q = 5, 7, 11, 9, 25, 15, 16, 20$ find a polynomial that is irreducible over $\mathbb{Q}$ and has a root equal to $\varepsilon_q := \cos(2\pi/q) + i \sin(2\pi/q)$.

(b) Same question with $\cos(2\pi/q)$ instead of ε_q.

First hints

8.3.2. It would be sufficient to show that the set of all numbers of the form $a \pm \sqrt{b}$, $a, b \in \mathbb{Q}$, is closed under addition, subtraction, multiplication, and division. However, this is obviously false: $(1 + \sqrt{2}) + (1 + \sqrt{3})$ cannot be represented as $a \pm \sqrt{b}$ where $a, b \in \mathbb{Q}$ (prove it!).

8.3.4. (a) If the polynomial $x^2 - r^2$ factors over $\mathbb{Q}$, then it has a rational root. This is a contradiction.

(b) If $b \neq 0$, then $r = -a/b \in \mathbb{Q}$, which is impossible. Hence $b = 0$, and thus $a = 0$.

(c) Consider the remainder upon dividing our polynomial by $x^2 - r^2$:[7]

$$P(x) = (x^2 - r^2)Q(x) + mx + n.$$

Substitute $x = r$. By the linear independence lemma (see (b)) the remainder is zero.

(d) By (c), if $R^2 = r^2$, then R is a root of the polynomial.

Sketch of alternative solution. The mapping $u \mapsto \overline{u}$ of the set $\mathbb{Q}[r] := \{a + br : a, b \in \mathbb{Q}\}$ into itself is well-defined by the formula $\overline{a + br} := a - br$. In addition, $\overline{u + v} = \overline{u} + \overline{v}$ and $\overline{u \cdot v} = \overline{u} \cdot \overline{v}$ for any $u, v \in \mathbb{Q}[\sqrt{2}]$.

(e) Let P be the given polynomial, and set $G(t) := P(a + bt)$. Then $G(r) = 0$. Hence by (d) we obtain $G(-r) = 0$.

(f) If $b = 0$, the assertion is proved. Otherwise, by (e), the polynomial has the roots $a \pm br$. These roots are distinct. Hence the third root is rational by Vieta's Theorem 3.6.5.

Suggestions, solutions, and answers

8.3.1. *Answers*: (a, b, c) Yes; (d, e, f, g) No.

(a, c) We have $\sqrt{3 + 2\sqrt{2}} = \sqrt[3]{7 + 5\sqrt{2}} = 1 + \sqrt{2}$.

(b) We have $\dfrac{1}{7 + 5\sqrt{2}} = \dfrac{7 - 5\sqrt{2}}{7^2 - 2 \cdot 5^2} = -7 + 5\sqrt{2}$.

(d) Assume that $\sqrt[3]{2}$ can be represented in this form. Then

$$2 = (\sqrt[3]{2})^3 = (a^3 + 3ab) + (3a^2 + b)\sqrt{b}.$$

Since $3a^2 + b \neq 0$, we have $\sqrt{b} \in \mathbb{Q}$. Thus $\sqrt[3]{2} \in \mathbb{Q}$, which is a contradiction.

Compare with problem 3.1.1 (h). The other proofs are similar to the proof of Theorem 8.3.5.

Yet another solution. Suppose that $\sqrt[3]{2}$ is representable. Then

$$1 = 1,$$
$$\sqrt[3]{2} = a + b\sqrt{2},$$
$$\sqrt[3]{4} = a' + b'\sqrt{2}$$

for some rational numbers a, b, a', and b'. The vectors $(1, 0)$, (a, b), and (a', b') are linearly dependent with rational coefficients, that is, there exist λ_0, λ_1, and λ_2, not all equal to zero, for which $\lambda_0(1, 0) + \lambda_1(a, b) + \lambda_2(a', b') = 0$. Then $\lambda_0 + \lambda_1\sqrt[3]{2} + \lambda_2\sqrt[3]{4} = 0$, which contradicts the linear independence lemma 8.3.18 (b).

(e) *Sketch of the first solution.* It is easier to prove a stronger assertion:

$$\sqrt[3]{2} \neq a + p\sqrt{b} + q\sqrt{c} + r\sqrt{bc} \quad \text{for any } a, b, c, p, q, r \in \mathbb{Q}.$$

---

[7]This is equivalent to "plugging in" $x^2 = r^2$.

It suffices to show that $\sqrt[3]{2} \neq u + v\sqrt{c}$ for any $u, v, c \in \mathbb{Q}[\sqrt{b}] := \{x + y\sqrt{b} \colon x, y \in \mathbb{Q}\}$. The idea of our proof is that numbers from $\mathbb{Q}[\sqrt{b}]$ (with fixed b) are "as good as" rational numbers. That is, the sum, the difference, the product, and the quotient of numbers from $\mathbb{Q}[\sqrt{b}]$ are also numbers from $\mathbb{Q}[\sqrt{b}]$ (or, in more advanced language, $\mathbb{Q}[\sqrt{b}]$ *is a field*). Then we can prove the assertion similarly to (d) (cf. problem 8.3.10 (a)).

Sketch of the second solution. Assume that $\sqrt{2} + \sqrt[3]{2} = a + \sqrt{b}$ for some $a, b \in \mathbb{Q}$. This number is a root of the polynomial $P(x) = ((x - \sqrt{2})^3 - 2)((x + \sqrt{2})^3 - 2)$ with rational coefficients. By 3.1.1 (h), $\sqrt{2} + \sqrt[3]{2} \notin \mathbb{Q}$. Hence $\sqrt{b} \notin \mathbb{Q}$. By the conjugation lemma 8.3.4 (e) for $r = \sqrt{b}$ we have $P(a - \sqrt{b}) = 0$. Since $\sqrt{b} \notin \mathbb{Q}$, the roots $a \pm \sqrt{b}$ are distinct. The polynomial P has only two real roots, namely $\sqrt{2} + \sqrt[3]{2}$ and $-\sqrt{2} + \sqrt[3]{2}$. Thus $a + \sqrt{b} = \sqrt{2} + \sqrt[3]{2}$ and $a - \sqrt{b} = -\sqrt{2} + \sqrt[3]{2}$. Therefore $\sqrt[3]{2} = a \in \mathbb{Q}$. This is a contradiction.

(f) The roots of the polynomial $P(x) = (x^2 - 2)^2 - 2$ are four numbers of the form $\pm\sqrt{2 \pm \sqrt{2}}$, where the signs need not agree. All these numbers are irrational. By Theorem 8.3.5, it is sufficient to prove that the polynomial P cannot be written as a product of two quadratic polynomials with rational coefficients. This irreducibility follows from the fact that the product of any two roots of P is irrational.

Sketch of another proof of irreducibility. By Gauss's lemma 8.4.10, it suffices to show that the polynomial does not decompose into a product of two quadratic trinomials with *integer* coefficients. Suppose the contrary. We can assume that the leading coefficients of these trinomials are both equal to 1. Since $P(0) = 2$, the constant term of one of the trinomials is even, and that of the other one is not. Then $x^4 = (x^2 + mx)(x^2 + nx + 1) \in \mathbb{Z}_2[x]$ for some $m, n \in \mathbb{Z}_2$. Comparing the coefficients at x and x^2, we obtain $m = 0$ and $0 = 1$, a contradiction. (Compare with the Eisenstein irreducibility criterion 8.4.9.)

(g) Use the hint for 3.1.1 (j).

8.3.2. Let α be a number obtained, starting with 1, using addition, subtraction, multiplication, division, and exactly one operation of taking the square root. Let this square root operation be $\sqrt{c}$, where $c \in \mathbb{Q}$. Then α can be written as $x = a_1 + a_2\sqrt{c}$ with $a_1, a_2 \in \mathbb{Q}$. Indeed, the set of numbers of such form is closed under all arithmetic operations; for division this can be proved using the formula $(a_1 + a_2\sqrt{c})(a_1 - a_2\sqrt{c}) = a_1^2 - a_2^2 c$. The proof follows since $a_1 + a_2\sqrt{c} = a_1 + b\sqrt{a_2^2 c}$. See also Lemma 8.4.6 (a).

8.3.3. *Answer*: The number is representable if and only if $n \in \{1, 2, 3, 4, 5, 6, 8, 10, 12\}$. Or, equivalently, $\varphi(n) \in \{1, 2, 4\}$. For $n \in \{15, 16, 20, 24\}$, see problem 8.1.3.

The case of $n = 9$: Suppose that $\cos(2\pi/9)$ is representable in this form. By formula 3.1.5 (e) for the cosine of a triple angle, $\cos(2\pi/9)$ is a root of the cubic equation $4x^3 - 3x = -\frac{1}{2}$. By Lemma 8.3.4 (f) this equation must have a rational root, a contradiction.

Another proof is analogous to Theorem 8.3.5.

The case of $n = 7$: (Here we follow I. Braude-Zolotarev.) The equality

$$\cos(2\pi/7) + \cos(4\pi/7) + \cos(6\pi/7) + \ldots + \cos(14\pi/7) = 0$$

implies that $\cos(2\pi/7) + \cos(4\pi/7) + \cos(6\pi/7) = -1/2$. Applying the formulas $\cos 2\alpha = 2\cos^2\alpha - 1$ and $\cos 3\alpha = 4\cos^3\alpha - 3\cos\alpha$ (see problem 3.1.5 (a, e)), we find that $\cos(2\pi/7)$ is a root of the equation $8t^3 + 4t^2 - 4t - 1 = 0$.

Substituting $u = 2t$, we get $u^3 + u^2 - 2u - 1 = 0$. This equation has no rational roots. Hence the same holds for $8t^3 + 4t^2 - 4t - 1 = 0$. Thus the polynomial $8t^3 + 4t^2 - 4t - 1 = 0$ is irreducible over $\mathbb{Q}$, and non-representability follows from Lemma 8.3.4 (f).

Instead of explicitly writing out the cubic equation with root $\cos(2\pi/7)$, one can see that the numbers $\cos(4\pi/7)$ and $\cos(6\pi/7)$ are also its roots. As in problem 3.1.4 (f) these numbers are irrational. Therefore, this cubic polynomial is irreducible over $\mathbb{Q}$.

Hint for alternative solution. Prove that ε_7 is not constructible by extracting only three square roots; compare with problems 8.3.9 and 8.3.14 (b).

8.3.5. Assume to the contrary that the given polynomial P has a root $x_0 = a \pm \sqrt{b}$, where $\sqrt{b} \notin \mathbb{Q}$. By the conjugation lemma 8.3.4 (e), the number $x_1 = a \mp \sqrt{b}$ is also a root of P. Since $\sqrt{b} \notin \mathbb{Q}$, we have $b \neq 0$. Then $x_0 \neq x_1$. Therefore P is divisible by $(x - a)^2 - b$. Since $\deg P > 2$, the polynomial P factors, a contradiction.

8.3.8. *Answers:*   (5) $x^4 + x^3 + x^2 + x + 1$;   (7) $x^6 + x^5 + \ldots + x + 1$;
(11) $x^{10} + x^9 + \ldots + x + 1$;   (9) $x^6 + x^3 + 1$;   (25) $x^{20} + x^{15} + x^{10} + x^5 + 1$;
(15) $(x^{15} - 1)(x - 1)/(x^5 - 1)(x^3 - 1)$;   (16) $x^8 + 1$.

(5) To prove irreducibility, apply the Eisenstein irreducibility criterion 8.4.9 to the polynomial $p(x + 1) = ((x + 1)^5 - 1)/x$ and use Gauss's lemma 8.4.10.

3.B. Multiple square root extractions (3*)

Here we develop ideas from subsection 3.A that are used to prove inconstructibility and insolvability.

8.3.9. Are there rational numbers a, b, c, d for which $\sqrt[3]{2}$ is equal to

(a) $a + b\sqrt[4]{2} + c\sqrt{2} + d\sqrt[4]{8}$;   (b) $\dfrac{a + \sqrt{b}}{c + \sqrt{b}}$;   (c) $a + \sqrt{b} + \sqrt{c}$;

(d) $a + \sqrt{b + \sqrt{c}}$;   (e) $a + \sqrt{b} + \sqrt{c} + \sqrt{d}$?

Lemma 8.3.10 (Linear independence). (a) If $a + b\sqrt[4]{2} + c\sqrt{2} + d\sqrt[4]{8} = 0$ for some $a, b, c, d \in \mathbb{Q}$, then $a = b = c = d = 0$.

(b) If $a + b\sqrt[4]{2} + c\sqrt{2} + d\sqrt[4]{8} = 0$ for some $a, b, c, d \in \mathbb{Q}[i] := \{x + iy : x, y \in \mathbb{Q}\}$, then $a = b = c = d = 0$.

Lemma 8.3.11 (Conjugation). (a) If $\sqrt[4]{2}$ is a root of a polynomial, then the following numbers are also roots of this polynomial: $-\sqrt[4]{2}$, $i\sqrt[4]{2}$, $-i\sqrt[4]{2}$.

(b) If $a, b, c, d \in \mathbb{Q}$ and a polynomial has the root $x_0 := a + b\sqrt[4]{2} + c\sqrt{2} + d\sqrt[4]{8}$, then the following numbers are also roots of this polynomial:

$$x_2 := a - b\sqrt[4]{2} + c\sqrt{2} - d\sqrt[4]{8},$$
$$x_1 := a - c\sqrt{2} + i\sqrt[4]{2}(b - d\sqrt{2}),$$
$$x_3 := a - c\sqrt{2} - i\sqrt[4]{2}(b - d\sqrt{2}).$$

(c) If a polynomial has the root $\sqrt{2} + \sqrt{3}$, then each of the four numbers $\pm\sqrt{2} \pm \sqrt{3}$ is a root of this polynomial.

Recall that definitions of real constructibility and constructibility are given in 1.B and 2.C respectively.[8]

8.3.12. If we remove the operation of division from the definition of constructibility but allow the use of all rational numbers, we get an equivalent definition.

Now you should be able to prove Theorem 8.1.2.

Hint. See the tower of extensions lemma 8.4.1 (a). For details see 4.D.

Theorem 8.3.13. (a) Some (or, equivalently, each) root of a cubic polynomial is constructible if and only if one of the roots of this polynomial is rational.

(b)* Some (or, equivalently, each) root of a fourth-degree polynomial is constructible if and only if its cubic resolvent (defined in the hint for problem 3.2.6 (b)) has a rational root.

8.3.14. The following numbers are not real constructible:

(a) $\cos(2\pi/9)$;

(b) $\cos(2\pi/7)$;

(c)* $\cos(2\pi/11)$.

---

[8] *Editor's note:* Here and in what follows complex constructible numbers will be called simply *constructible.*

8.3.15. Find a polynomial irreducible over $\mathbb{Q}$ with root

(a) $\sqrt{2} + \sqrt{3}$;   (b) $\sqrt{4 + 2\sqrt{3}}$;   (c) $\sqrt{2} + \sqrt{3} + \sqrt{5}$;

(d) $\sqrt{1 + \sqrt{3}}$;   (e) $\sqrt{2 + \sqrt{2 + \sqrt{5}}}$.

8.3.16. (a) If a polynomial P is irreducible over $\mathbb{Q}[\sqrt{2}] := \{x + y\sqrt{2} : x, y \in \mathbb{Q}\}$ and has a root of the form $a + \sqrt{b}$ where $a, b \in \mathbb{Q}[\sqrt{2}]$, then $\deg P \in \{1, 2\}$.

(b) If a polynomial P is irreducible over $\mathbb{Q}$ and has a root of the form $a + \sqrt{b} + \sqrt{c}$ where $a, b, c \in \mathbb{Q}$, then $\deg P \in \{1, 2, 4\}$.

(c) If a polynomial P is irreducible over $\mathbb{Q}$ and has a root of the form $\sqrt{a} + \sqrt{b + \sqrt{c}}$ where $a, b, c \in \mathbb{Q}$, then $\deg P \in \{1, 2, 4, 8\}$.

(d) If a polynomial P is irreducible over $\mathbb{Q}$ and has a constructible root, then $\deg P$ is a power of 2.

The proofs are similar to those of problems 8.3.15 (d, e). See the proof of 8.4.7 for details.

Suggestions, solutions, and answers

8.3.9. *Answers*: No. See 8.3.1 (e, g).

(a) *First solution.* Suppose it is expressible in the given form. By the conjugation lemma 8.3.11 (b), the polynomial $x^3 - 2$ has roots x_0 and x_2 introduced in the statement of the lemma. Since neither of them is rational, the equality $b = d = 0$ is impossible. So, by the linear independence lemma 8.3.10 (a), these roots are distinct, a contradiction.

Second solution. Suppose to the contrary that it is expressible in the given form. By the conjugation lemma 8.3.11 (b), the polynomial $x^3 - 2$ has the roots x_1, x_2, x_3, x_4 introduced in the statement of the lemma. Since none of them is rational, these roots are pairwise distinct, a contradiction.

(b) Multiply by the conjugate.

8.3.10. (a) *First solution.* Rewrite the condition in the form $(a + c\sqrt{2}) + (b + d\sqrt{2})\sqrt[4]{2} = 0$. Since $b + d\sqrt{2} \neq 0$, we have $-\sqrt[4]{2} = \frac{a + c\sqrt{2}}{b + d\sqrt{2}} = A + B\sqrt{2}$ for some $A, B \in \mathbb{Q}$. Squaring yields $A^2 + 2B^2 = 0$, a contradiction.

Second solution. Considering the complex roots of the polynomial $v^4 - 2$, we see that it is irreducible over $\mathbb{Q}$. Therefore it cannot have a common root with the polynomial $a + bx + cx^2 + dx^3$, which is of at most third degree.

(b) Prove the assertion separately for the real and imaginary parts.

8.3.11. (a) Divide the polynomial by $x^4 - 2$ and consider its remainder. After substituting $x = \sqrt[4]{2}$, the linear independence lemma 8.3.10 (a) implies that the remainder is zero. Therefore, if $r^2 = 2$, then $a + br + cr^2 + dr^3$ is a root of the original polynomial.

(b) Substitute $x = a + bt + ct^2 + dt^3$ into the polynomial and use part (a).

8.3.12. The assertion can be proved similarly to Lemma 8.4.1 (a).

8.3.13. (a) The "if" part is easy. To prove "only if," we suppose to the contrary that at least one root is constructible. Then for each constructible root z there exists an extension tower as asserted by Lemma 8.4.1 (a). We can assume that $z \in F_s - F_{s-1}$. Take the root s with minimal $s = s(z)$ (among all constructible roots z). Conversely, assume that the cubic equation does not have rational roots. Then $s \geq 2$. Let $F := F_{s-1}$. Then

$$z = a + b\sqrt{m} \quad \text{for some} \quad a, b, m \in F, \quad m > 0, \quad \sqrt{m} \notin F, \quad b \neq 0.$$

By the conjugation lemma 8.4.8 (a), $\overline{z} := a - b\sqrt{m}$ is also a root of the equation. Since $b \neq 0$, we have $a - b\sqrt{m} \neq a + b\sqrt{m}$. Let u be the third root of the equation (maybe $u \in \{z, \overline{z}\}$). By Vieta's Theorem 3.6.5,

$$\mathbb{Q} \ni u + z + \overline{z} = u + (a + b\sqrt{m}) + (a - b\sqrt{m}) = u + 2a.$$

Therefore $u \in F$. Thus, for the root u, there exist extension towers of lower height than those for z. This is a contradiction.

8.3.14. (a) See 8.3.13 (a).
   (b) Let $\varepsilon := \varepsilon^7$. Since $\varepsilon \neq 1$, ε is a root of the 6th-degree equation $\varepsilon^6 + \varepsilon^5 + \varepsilon^4 + \varepsilon^3 + \varepsilon^2 + \varepsilon + 1 = 0$. Below is one way to solve nth-degree polynomial equations that have equal coefficients at the kth and $(n - k)$th degrees.
   Divide both parts of the equation by ε^3, and make the substitution $t := \varepsilon + \varepsilon^{-1}$; then

$$\varepsilon^2 + \varepsilon^{-2} = t^2 - 2 \quad \text{and} \quad \varepsilon^3 + \varepsilon^{-3} = t(\varepsilon^2 + \varepsilon^{-2} - 1),$$

leading to

$$t(t^2 - 3) + (t^2 - 2) + t + 1 = t^3 + t^2 - 2t - 1 = 0.$$

By the rational roots theorem, there are no rational roots. By Theorem 8.3.13 (a), $t = \varepsilon + \varepsilon^{-1}$ is not constructible, which implies that ε is also non-constructible (why?).

8.3.15. *Answers*:
   (a) $P(x) := ((x - \sqrt{3})^2 - 2)((x + \sqrt{3})^2 - 2) = (x^2 + 1)^2 - 12x^2$;
   (b) $(x^2 - 3)^2 - 1$;   (c) $P(x - \sqrt{5})P(x + \sqrt{5})$;
   (d) $(x^2 - 1)^2 - 3$;   (e) $((x^2 - 2)^2 - 2)^2 - 5$.
   (a) To prove irreducibility, use the conjugation lemma, which implies that each of the four numbers $\pm\sqrt{2}\pm\sqrt{3}$ is a root of the desired polynomial. Then prove and use the fact that $\sqrt{3} \notin \mathbb{Q}[\sqrt{2}]$.
   (b) Note that $\sqrt{4 + 2\sqrt{3}} = 1 + \sqrt{3}$.

(c) The proof is similar to that of (a). Prove and use the fact that $\sqrt{5} \notin \mathbb{Q}[\sqrt{2}][\sqrt{3}]$. Similarly to problem 8.3.9, it follows from the statement that if the polynomial $x^2 - 5$ has the root $a + b\sqrt{3}$ where $a, b \in \mathbb{Q}[\sqrt{2}]$, then it also has the root $a - b\sqrt{3}$.

(d) Similarly to (a) (or similarly to the following), the *monic* (leading coefficient is 1) irreducible polynomial with root $\sqrt{1 + \sqrt{3}}$ and coefficients in $\mathbb{Q}[\sqrt{3}]$ is $x^2 - 1 - \sqrt{3}$. Therefore any irreducible polynomial P with coefficients in $\mathbb{Q}$ and root $\sqrt{1 + \sqrt{3}}$ is divisible by $x^2 - 1 - \sqrt{3}$ in $\mathbb{Q}[\sqrt{3}]$. Applying conjugation with respect to $\mathbb{Q} \subset \mathbb{Q}[\sqrt{3}]$, we conclude that P over $\mathbb{Q}[\sqrt{3}]$ is divisible by $x^2 - 1 + \sqrt{3}$. Since the polynomials $x^2 - 2 - \sqrt{3}$ and $x^2 - 2 + \sqrt{3}$ are coprime in $\mathbb{Q}[\sqrt{3}]$, P over $\mathbb{Q}[\sqrt{3}]$ is divisible by their product.

(e) Verify that $x^2 = 2 + \sqrt{5}$ has no root in $\mathbb{Q}[\sqrt{5}]$:

If the equation $x^2 = 2 + \sqrt{2 + \sqrt{5}}$ were solvable in $\mathbb{Q}[\sqrt{2 + \sqrt{5}}]$, then conjugation in $\mathbb{Q}[\sqrt{2 + \sqrt{5}}] = \mathbb{Q}[\sqrt{5}, \sqrt{2 + \sqrt{5}}]$ with respect to $\mathbb{Q}[\sqrt{5}]$ would yield $\overline{x}^2 = 2 - \sqrt{2 + \sqrt{5}} < 0$. This contradiction proves that the equation $x^2 = 2 + \sqrt{2 + \sqrt{5}}$ has no roots in $\mathbb{Q}[\sqrt{2 + \sqrt{5}}]$.

Similarly to (d), we see that the given irreducible polynomial with root $\sqrt{2 + \sqrt{2 + \sqrt{5}}}$ and coefficients in $\mathbb{Q}[\sqrt{2 + \sqrt{5}}]$ is equal to $x^2 - 2 - \sqrt{2 + \sqrt{5}}$.

In $\mathbb{Q}[\sqrt{5}]$ it is equal to $(x^2 - 2)^2 - 2 - \sqrt{5}$.

In $\mathbb{Q}$ it is equal to $((x^2 - 2)^2 - 2)^2 - 5$.

3.C. Representing a number using only one cube root (2)

Here we develop the ideas from 3.A (in a different direction than in 3.B).

8.3.17. Which of the following numbers can be represented in the form $a + b\sqrt[3]{2} + c\sqrt[3]{4}$ with $a, b, c \in \mathbb{Q}$?

(a) $\sqrt{3}$;   (b) $\dfrac{1}{1 + 5\sqrt[3]{2} + \sqrt[3]{4}}$;   (c) $\cos(2\pi/9)$;   (d) $\sqrt[5]{3}$;   (e) $\sqrt[3]{3}$;

(f) the largest real root of $x^3 - 4x + 2 = 0$;

(g)* the unique real root of $x^3 - 6x - 6 = 0$;

(h)* the unique real root of $x^3 - 9x - 12 = 0$.

Lemma 8.3.18. Assume that $r \in \mathbb{R} - \mathbb{Q}$ and $r^3 \in \mathbb{Q}$.

(a) **Irreducibility.** The polynomial $x^3 - r^3$ is irreducible over $\mathbb{Q}$.

(b) **Linear independence.** If $a + br + cr^2 = 0$ with $a, b, c \in \mathbb{Q}$, then $a = b = c = 0$.

(b′) **Linear independence over** $\mathbb{Q}[\varepsilon_3]$. If

$$k, \ell, m \in \mathbb{Q}[\varepsilon_3] := \{u + v\varepsilon_3 : u, v \in \mathbb{Q}\}$$

and $k + \ell r + mr^2 = 0$, then $k = \ell = m = 0$.

(c) If r is a root of a polynomial, then this polynomial is divisible by $x^3 - r^3$.

(d) **Conjugation.** If r is a root of a polynomial, then the numbers $\varepsilon_3 r$ and $\varepsilon_3^2 r$ are also its roots.

(e) **Conjugation.** If $a, b, c \in \mathbb{Q}$ and a polynomial has root $x_0 := a + br + cr^2$, then the numbers

$$x_1 := a + b\varepsilon_3 r + c\varepsilon_3^2 r^2 \quad \text{and} \quad x_2 := a + b\varepsilon_3^2 r + c\varepsilon_3 r^2$$

are also its roots.

(f) **Rationality.** If $a, b, c \in \mathbb{Q}$, then $a + br + cr^2$ is a root of some cubic polynomial.

Theorem 8.3.19. If a polynomial is irreducible over $\mathbb{Q}$ and has a root $a + br + cr^2$ for some $r \in \mathbb{R} - \mathbb{Q}$ and $a, b, c, r^3 \in \mathbb{Q}$, then this polynomial is cubic and has exactly one real root.

Lemma 8.3.20 (Extension). A number expressible in real radicals with only one extraction of a cube root can be represented in the form $a + br + cr^2$ where $r \in \mathbb{R}$ and $a, b, c, r^3 \in \mathbb{Q}$.

Suggestions, solutions, and answers

8.3.17. *Answers*: (a, c, d, e, f, h) No; (b, g) Yes.

Let $r := \sqrt[3]{2}$.

(a) Assume that $\sqrt{3}$ is representable in this form.

First solution. Then

$$3 = (a^2 + 4bc) + (2ab + 2c^2)\sqrt[3]{2} + (2ac + b^2)\sqrt[3]{4}.$$

Since the polynomial $x^3 - 2$ has no rational roots, it is irreducible over $\mathbb{Q}$. Thus, $2ab + 2c^2 = 2ac + b^2 = 0$ (cf. 8.3.18 (b)). So we have $b^3 = -2abc = 2c^3$. Hence either $b = c = 0$ or $\sqrt[3]{2} = b/c$. Both cases are impossible.

Second solution. Let $P(x) := x^2 - 3$. By the conjugation lemma 8.3.18 (e), P has three roots x_0, x_1, x_2 as defined in the statement of the lemma. Since none of them is rational, the equality $b = c = 0$ does not hold. By the linear independence lemma over $\mathbb{Q}[\varepsilon_3]$ 8.3.18 (b'), the three roots are distinct, a contradiction.

(b) We have $(1 + 5\sqrt[3]{2} + \sqrt[3]{4})(3 + \sqrt[3]{2} - 8\sqrt[3]{4}) = -75$. (This equality can be easily obtained by the undetermined coefficients method or by applying the Euclidean algorithm to $x^3 - 2$ and $x^2 + 5x + 1$; see the solution to 8.3.20.) Therefore,

$$\frac{1}{1 + 5\sqrt[3]{2} + \sqrt[3]{4}} = -\frac{1}{25} - \frac{1}{75} \cdot \sqrt[3]{2} + \frac{8}{75} \cdot (\sqrt[3]{2})^2.$$

(c) Assume that $\cos(2\pi/9)$ is representable in this form. This number is a root of the equation $4x^3 - 3x = -\frac{1}{2}$. Its other two real roots are $\cos(8\pi/9)$ and $\cos(4\pi/9)$.

First solution. From the conjugation lemma 8.3.18 (e), the polynomial $8x^3 - 6x - 1$ has a root $a + br\varepsilon_3 + cr^2\varepsilon_3^2$. Since it is real, its imaginary part is equal to zero. Thus, $br - cr^2 = 0$. Since $r \notin \mathbb{Q}$, we have $b = c = 0$, which contradicts the fact that $8x^3 - 6x - 1$ has no rational roots.

Second solution. Following the second solution to (a), note that $P(x) := 8x^3 - 6x - 1$ has three distinct roots x_0, x_1, x_2. Since $\overline{\varepsilon_3} = \varepsilon_3^2$, we have $\overline{x_2} = x_1$. Thus, x_1 and x_2 cannot be both real and distinct, a contradiction.

(d) Assume that $\sqrt[5]{3}$ is representable in this form. By the rationality lemma 8.3.18 (f), $\sqrt[5]{3}$ is a root of a cubic polynomial. This contradicts the irreducibility of the polynomial $x^5 - 3$ over $\mathbb{Q}$.

(e) As with (a) and (c), by the conjugation lemma 8.3.18 (e), it follows that the polynomial $x^3 - 3$ has three roots x_0, x_1, x_2 as defined in the statement of the lemma. Thus, $(a + br + cr^2)\varepsilon_3^s = a + br\varepsilon_3 + cr^2\varepsilon_3^2$ for some $s \in \{1, 2\}$. By the linear independence lemma over $\mathbb{Q}[\varepsilon_3]$ 8.3.18 (b'), we conclude that $a = 0$ and $bc = 0$. Hence either $\sqrt[3]{3} = br$ or $\sqrt[3]{3} = cr^2$, a contradiction.

(f) See (c).

(g) One root of this equation is $\sqrt[3]{2} + \sqrt[3]{4}$.

(h) The only real root of this equation is $\sqrt[3]{3} + \sqrt[3]{9}$. Assume that this number is representable in the required form. Repeat the second solution of (a) for $P(x) := x^3 - 9x - 12$. We obtain that x_0, x_1, x_2 are all roots of P.

On the other hand, by the theorem formulated in the solution to 3.2.4 (b), all roots of this equation are

$$y_0 := \sqrt[3]{3} + \sqrt[3]{9}, \quad y_1 := \sqrt[3]{3}\varepsilon_3 + \sqrt[3]{9}\varepsilon_3^2, \quad y_2 := \sqrt[3]{3}\varepsilon_3^2 + \sqrt[3]{9}\varepsilon_3.$$

Since the equation has exactly one real root, we must have $x_0 = y_0$. Then either $x_1 = y_1$ and $x_2 = y_2$, or $x_2 = y_1$ and $x_1 = y_2$.

Define $R(x) := \sqrt[3]{3}x + \sqrt[3]{9}x^2$ and $S(x) := a + brx + cr^2x^2$ or $S(x) := a + brx^2 + cr^2x$ in the first and second cases, respectively. Then the polynomial $R(x) - S(x)$ has three distinct roots 1, ε_3, and ε_3^2. But the degree of this polynomial is at most 2. Thus $R = S$, and either $\sqrt[3]{3} = br$ or $\sqrt[3]{3} = cr^2$, a contradiction.

8.3.18. (a) Suppose that $x^3 - r^3$ is reducible over $\mathbb{Q}$. Then it has a rational root, a contradiction.

(b) Assume not. Divide $x^3 - r^3$ by $a + bx + cx^2$ and consider the remainder. By (a), the remainder is nonzero. Both $x^3 - r^3$ and $a + bx + cx^2$ have a root $x = r$. Hence the remainder has the root $x = r$. Thus, the remainder has an irrational root. This is impossible because the remainder has degree 1.

(b') Consider the real and imaginary parts separately.

Remark. This assertion is equivalent to the irreducibility of $x^3 - r^3$ over $\mathbb{Q}[\varepsilon_3]$. If $x^3 - r^3$ is irreducible over $\mathbb{Q}[\varepsilon_3]$, then $k + lx + mx^2 \in \mathbb{Q}[\varepsilon_3][x]$ cannot have the root r. If $x^3 - r^3$ factors over $\mathbb{Q}[\varepsilon_3]$, then an examination of the factors implies the linear dependence of $1, r, r^2$ over $\mathbb{Q}[\varepsilon_3]$.

(c) Divide $x^3 - r^3$ and consider the remainder. Taking $x = r$ and applying the linear independence lemma 8.3.18 (b), we see that the remainder is zero.

(d) By (c), if $R^3 = r^3$, then R is a root of the polynomial.

(e) Let P be the given polynomial, and set $G(t) := P(a + bt + ct^2)$. Then $G(r) = 0$. Hence by (d) we have $G(r\varepsilon_3) = 0 = G(r\varepsilon_3^2)$.

(f) *First solution.* Taking $x = y + a$, we see that it suffices to prove the assertion for $a = 0$. Note that $t = br + cr^2$ satisfies $t^3 = b^3 r^3 + c^3 r^6 + 3bcr^3 t$.

In other words, since $u^3 + v^3 + w^3 - 3uvw$ is divisible by $u + v + w$ (see problem 3.2.3 (a), the number $a + br + cr^2$ is a root of the polynomial

$$(x - a)^3 - 3bcr^3(x - a) - b^3 r^3 - c^3 r^6.$$

Second solution. Let $x_0 := a + br + cr^2$. Expand the numbers x_0^k, $k = 0, 1, 2, 3$, as polynomials in r:

$$x_0^k = a_k + b_k r + c_k r^2.$$

It suffices to find numbers $\lambda_0, \lambda_1, \lambda_2, \lambda_3 \in \mathbb{Q}$, not all zero, such that $\lambda_0 + \lambda_1 x_0 + \lambda_2 x_0^2 + \lambda_3 x_0^3 = 0$. These numbers must satisfy the system of equations

$$\begin{cases} \lambda_0 a_0 + \ldots + \lambda_3 a_3 = 0, \\ \lambda_0 b_0 + \ldots + \lambda_3 b_3 = 0, \\ \lambda_0 c_0 + \ldots + \lambda_3 c_3 = 0. \end{cases}$$

It is known that a homogeneous (i.e., with zeros on the right-hand sides) system of linear equations with rational coefficients where the number of equations is smaller than the number of variables has a nontrivial rational solution. Hence, the required numbers exist.

The resulting polynomial has degree three by Lemma 8.3.18 (e, b').

Third solution. Let $A(x) := a + bx + cx^2$. The product $(x - A(t_0))(x - A(t_1))(x - A(t_2))$ is a symmetric polynomial in t_0, t_1, t_2. Hence this product is a polynomial in x and the elementary symmetric polynomials in t_0, t_1, t_2. The values of these elementary symmetric polynomials at $t_k = r\varepsilon_3^k$ ($k = 0, 1, 2$) are the coefficients of the polynomial $x^3 - r^3$ and hence are rational. So the product above is the required polynomial.

8.3.19. By the rationality lemma 8.3.18 (f) there exists a cubic polynomial having $a + br + cr^2$ as a root. Since the given polynomial P is irreducible over $\mathbb{Q}$ and has the same root, we conclude that $\deg P \leq 3$. By the conjugation lemma 8.3.18 (e), P has three roots x_0, x_1, x_2 as defined in the statement of the lemma. Since P is irreducible over $\mathbb{Q}$, none of its roots is rational. So the equality $b = c = 0$ is impossible. By the linear independence lemma over $\mathbb{Q}[\varepsilon_3]$ 8.3.18 (b'), x_0, x_1, x_2 are distinct. Hence $\deg P = 3$.

Since $\overline{\varepsilon_3^k} = \varepsilon_3^{-k}$, we have $\overline{x_2} = x_1$. Thus x_2 and x_1 cannot be real and distinct. So $x_2, x_1 \in \mathbb{C} - \mathbb{R}$, implying that P has a unique real root.

8.3.20. Assume that after extracting a cube root we obtain r. If $|r| \in \mathbb{Q}$, the statement is trivial. If $|r| \notin \mathbb{Q}$, then the polynomial $x^3 - r^3$ is irreducible over $\mathbb{Q}$.

It suffices to prove that $\frac{1}{a+br+cr^2} = h(r)$ for some polynomial h. By the irreducibility lemma, the polynomial $x^3 - r^3$ is irreducible over $\mathbb{Q}$. Hence it is coprime with $a + bx + cx^2$. Therefore, there exist polynomials g and h such that $h(x)(a + bx + cx^2) + g(x)(x^3 - r^3) = 1$. Then h is the required polynomial.

3.D. Representing a number using only one root of prime order (3*)

This subsection expands upon the ideas of 3.C.

8.3.21. Which of the following numbers can be represented in the form
$$a_0 + a_1 \sqrt[7]{2} + a_2 \sqrt[7]{2^2} + \cdots + a_6 \sqrt[7]{2^6}$$
with $a_0, a_1, a_2, \ldots, a_6 \in \mathbb{Q}$?
   (a) $\sqrt{3}$;   (b) $\cos \frac{2\pi}{21}$;   (c) $\sqrt[11]{3}$;   (d) $\sqrt[7]{3}$;
   (e) a root of the polynomial $x^7 - 4x + 2$.

Answers: None. The arguments are similar to those used in 8.3.17. Use the lemmas stated below.

Lemma 8.3.22. Let q be a prime number and let $r \in \mathbb{R} - \mathbb{Q}$ such that $r^q \in \mathbb{Q}$.
   (a) **Irreducibility.** The polynomial $x^q - r^q$ is irreducible over $\mathbb{Q}$.
   (b) **Linear independence.** If r is a root of a polynomial A of degree less than q, then $A = 0$.
   (c) **Conjugation.** If r is a root of a polynomial, then all the numbers $r\varepsilon_q^k$, $k = 1, 2, 3, \ldots, q - 1$, are also roots of this polynomial.
   (d) **Rationality.** If A is a polynomial, then the number $A(r)$ is a root of some nonzero polynomial of degree at most q.

8.3.23. Define
$$\mathbb{Q}[\varepsilon_q] := \{a_0 + a_1\varepsilon_q + a_2\varepsilon_q^2 + \ldots + a_{q-2}\varepsilon_q^{q-2} \ : \ a_0, \ldots, a_{q-2} \in \mathbb{Q}\}.$$

Let q be a prime number and let $r \in \mathbb{C} - \mathbb{Q}[\varepsilon_q]$ such that $r^q \in \mathbb{Q}[\varepsilon_q]$.
   (a) Prove that $x^q - r^q$ is irreducible over $\mathbb{Q}[\varepsilon_q]$.
   (b, c) Prove the analogues of (b, c) in the lemma above for polynomials with coefficients in $\mathbb{Q}[\varepsilon_q]$.

Lemma 8.3.24.* Let q be a prime number and let $r \in \mathbb{R} - \mathbb{Q}$ such that $r^q \in \mathbb{Q}$.

(a) **Irreducibility over** $\mathbb{Q}[\varepsilon_q]$. The polynomial $x^q - r^q$ is irreducible over $\mathbb{Q}[\varepsilon_q]$.

(b) **Linear independence over** $\mathbb{Q}[\varepsilon_q]$. If A is a polynomial of degree less than q with coefficients in $\mathbb{Q}[\varepsilon_q]$ and $A(r) = 0$, then $A = 0$.

Theorem 8.3.25. Suppose that a polynomial B is irreducible over $\mathbb{Q}$ and has an irrational root $A(r)$, where A is a polynomial and $r \in \mathbb{R}$ is such that $r^q \in \mathbb{Q}$ for some prime q. Then B has degree q and, if $q \neq 2$, has no other real roots.

The proof is analogous to the proofs of Theorems 8.3.5 and 8.3.19 and to the solutions of 8.3.21 (a, b, c). Apply the conjugation lemma 8.3.22 (c), the rationality lemma 8.3.22 (d), and the linear independence lemma over $\mathbb{Q}[\varepsilon_q]$ 8.3.24 (b).

Note that the analogue of Theorem 8.3.25 fails if we replace the condition that q is a prime by the condition $r^2, \ldots, r^{q-1} \notin \mathbb{Q}$. (For example, let $q = 6$ and $r = \sqrt[6]{2}$.) Then the number $A(r) = r^3$ is a root of $x^2 - 2$.

Lemma 8.3.26 (Extension). If a number is expressible in real radicals with only one root extraction, then it equals $A(r)$ for some $r \in \mathbb{R}$, $q \in \mathbb{Z}$ with $r^q \in \mathbb{Q}$, and $A \in \mathbb{Q}[x]$.

The proof is similar to the proof of the extension lemma 8.3.20.

8.3.27. (a–d) Prove the analogues of the assertions in 8.3.22 with $\mathbb{Q}$ replaced by any set $F \subset \mathbb{R}$ which is closed under the operations of addition, subtraction, multiplication, and division by a nonzero number (and with polynomials over $\mathbb{Q}$ replaced by polynomials over F).

Suggestions, solutions, and answers

8.3.21. Set $r := \sqrt[7]{2}$ and $A(x) := a_0 + a_1 x + a_2 x^2 + \ldots + a_6 x^6$.

(a) Assume that $\sqrt{3}$ is representable in this form. By the conjugation lemma 8.3.22 (c), the polynomial $x^2 - 3$ has roots $A(r\varepsilon_7^k)$ for $k = 0, 1, 2, \ldots, 6$. Since this polynomial has no rational roots, the linear independence lemma over $\mathbb{Q}[\varepsilon_q]$ 8.3.24 (b) implies that these roots are distinct, a contradiction.

(b) Assume that $\cos \frac{2\pi}{21}$ is representable in this form.

First solution. Similarly to (a), the given polynomial P has pairwise distinct roots $x_k := A(r\varepsilon_7^k)$ for $k = 0, 1, 2, \ldots, 6$. Since $P(0) > 0$, $P(1) < 0$, and $P(2) > 0$, P has a real root x_k distinct from x_0. Since $\overline{\varepsilon_7^k} = \varepsilon_7^{-k}$, we have $x_k = \overline{x_k} = x_{7-k}$, a contradiction.

Second solution. Define P to be the polynomial such that $\cos 7x = P(\cos x)$ (see 3.1.6(a)). The roots of the polynomial $2P(x) + 1$ are real numbers $y_k = \cos \dfrac{2(3k+1)\pi}{21}$ with $k = 0, \ldots, 6$. One of them, namely $y_2 = -1/2$, is rational.

Next, we prove that y_0 is irrational. If it is not, the equality $\varepsilon_{21}^2 - 2y_0\varepsilon_{21} + 1 = 0$ implies that $\varepsilon_{21} = a + i\sqrt{b}$ for some $a, b \in \mathbb{Q}$. Then the number $\varepsilon_7 = \varepsilon_{21}^3$ also has this form. But ε_7 is a root of the irreducible[9] polynomial $1 + x + \cdots + x^6$, which contradicts the analogue of Theorem 8.3.5 for numbers of the form $a + i\sqrt{b}$.

Consequently, y_0 is an irrational root of the sixth-degree polynomial $\frac{2P(x)+1}{2x+1}$. The conjugation lemma 8.3.22 (c) and linear independence lemma over $\mathbb{Q}[\varepsilon_q]$ 8.3.24 (b) imply that this polynomial has seven distinct roots, which is impossible.

(c) Assume that $\sqrt[11]{3}$ is representable in this form. Then by the rationality lemma 8.3.22 (d), there exists a nonzero polynomial of degree at most 7 having $\sqrt[11]{3}$ as a root. This contradicts the irreducibility of the polynomial $x^{11} - 3$ over $\mathbb{Q}$.

(d) Assume that $\sqrt[7]{3}$ is representable in this form. Similarly to (a), all complex roots of the polynomial $x^7 - 3$ are $A(r\varepsilon_7^k)$ for $k = 0, 1, 2, \ldots, 6$. Therefore, $A(r)\varepsilon_7^s = A(r\varepsilon_7)$ for some $s \in \{1, 2, 3, 4, 5, 6\}$. By the linear independence lemma over $\mathbb{Q}[\varepsilon_q]$ 8.3.24 (b), we have $a_k = 0$ for every $k \neq s$. Therefore, $\sqrt[7]{3} = a_s r^s$, a contradiction.

(e) Assume that one of the roots is representable in this form. The given polynomial P has no rational roots. The conjugation lemma 8.3.22 (c) and linear independence lemma over $\mathbb{Q}[\varepsilon_q]$ 8.3.24 (b) imply that P has pairwise distinct roots $x_k := A(r\varepsilon_7^k)$ for $k = 0, 1, 2, \ldots, 6$. Since $P(0) > 0$, $P(1) < 0$, and $P(2) > 0$, the polynomial P has a real root x_k distinct from x_0. From the equality $\overline{\varepsilon_7^k} = \varepsilon_7^{-k}$ it follows that $x_k = \overline{x_k} = x_{7-k}$, a contradiction.

8.3.22. (a) All roots of the polynomial $x^q - r^q$ are $r, r\varepsilon_q, r\varepsilon_q^2, \ldots, r\varepsilon_q^{q-1}$. Assume that $x^q - r^q$ is reducible over $\mathbb{Q}$, which means that there is a polynomial with rational coefficients of degree $k < q$ which divides it. The value of the constant term of this polynomial is rational and equals the product of the absolute values of k of the roots, $0 < k < q$. Therefore, $r^k \in \mathbb{Q}$. Since q is prime, we have $kx + qy = 1$ for some integers x and y. Thus, $r = (r^k)^x (r^q)^y \in \mathbb{Q}$. This is a contradiction.

(b) Assume not. Consider the polynomial $A(x)$ of lowest degree for which the statement is false. Let $R(x)$ be the remainder when $x^q - r^q$ is divided by $A(x)$. Then $\deg R < \deg A$, $R(r) = 0$, and $R(x) \neq 0$ by (a). This contradicts the choice of A.

---

[9]The irreducibility of the polynomial $g(x) = 1 + x + \ldots + x^6$ can be proved, e.g., by applying the Eisenstein criterion 8.4.9 to the polynomial $g(x + 1)$. However, in this particular case it suffices to prove that g has no divisors of degree 1 or 2 with rational coefficients.

(c) The solution is analogous to the solutions of 8.3.4 (c, d), 8.3.18 (d), and 8.3.11 (a). Use (b).

(d) The proofs repeat the second and third proofs of the rationality lemma 8.3.18 (f). It is only necessary to replace 3 by q and 2 by $q - 1$ throughout the proofs (for example, in the second line of the second proof put $k = 0, 1, 2, \ldots, q$). Compare with the proof of the decomposition lemma in 4.G.

8.3.23. (a) Assume that the polynomial can be factored over $\mathbb{Q}[\varepsilon_q]$. The constant term of any of these factors lies in $\mathbb{Q}[\varepsilon_q]$ and equals $\pm r^k \varepsilon_q^m$ for some m. Then $r^k \in \mathbb{Q}[\varepsilon_q]$. Following the proof of Lemma 8.3.22 (a) we see that $r \in \mathbb{Q}[\varepsilon_q]$, a contradiction.

Parts (b, c) are deduced from (a) analogously to 8.3.22 (b, c).

8.3.24. (a) Suppose that the polynomial is reducible. Following the proof of irreducibility over $\mathbb{Q}[\varepsilon_q]$, Lemma 8.3.23 (a), we have $r \in \mathbb{Q}[\varepsilon_q]$. Thus, $r^2, r^3, \ldots, r^{q-1} \in \mathbb{Q}[\varepsilon_q]$.

Now we show that r is a root of a polynomial of degree at most $q - 1$, which will contradict the irreducibility of $x^q - r^q$ over $\mathbb{Q}$. Expand the numbers r^k as polynomials in ε_q for $k = 0, 1, \ldots, q - 1$:

$$r^k = a_{k,0} + a_{k,1}\varepsilon_q + \ldots + a_{k,q-2}\varepsilon_q^{q-2}.$$

It suffices to find numbers $\lambda_0, \lambda_1, \ldots, \lambda_{q-1} \in \mathbb{Q}$, not all of them zero, such that

$$\lambda_0 a_{0,m} + \ldots + \lambda_{q-1} a_{q-1,m} = 0 \quad \text{for every } m = 0, 1, \ldots, q - 2.$$

Such numbers exist analogously to the corresponding assertion in the second proof of the rationality lemma 8.3.18 (f). (In other words, take an array of size $q \times (q - 1)$ formed by the rational numbers a_{kl}. By several operations of adding a row multiplied by a rational number to another row, one can get an array containing a zero row.)

Part (b) follows from (a).

8.3.25. Assume not. Let P be the given polynomial. The assumption $q < \deg P$ contradicts the rationality lemma 8.3.22 (d). If $q \geq \deg P$, then by the conjugation lemma 8.3.22 (c) and the linear independence lemma over $\mathbb{Q}[\varepsilon_q]$ 8.3.24 (b), the polynomial P has pairwise distinct roots $x_k = A(r\varepsilon_q^k)$ for $k = 0, 1, \ldots, q - 1$. If $q > \deg P$ we get a contradiction. If $q = \deg P$ the conditions $q \neq 2$ and $\overline{x_k} = x_{q-k} \neq x_k$ imply the uniqueness of the real root.

3.E. There is only one way to solve a quadratic equation (2)

In this and the following subsections, equality signs involving polynomials f (or f_j) mean equality of polynomials coefficientwise.

The systems of equations studied here and in the following subsection arise when solving equations in radicals ("using one radical"); see Remark 8.2.1 (d).

8.3.28. (a, b) Solve the following systems of equations, where $f(x, y)$, $p(u, v)$, and $q(u, v, w)$ are polynomials with real coefficients.

(a) $\begin{cases} f^2(x, y) = p(x + y, xy), \\ x = q(x + y, xy, f(x, y)). \end{cases}$

(b) $\begin{cases} f^k(x, y) = p(x + y, xy), \\ x = q(x + y, xy, f(x, y)), \end{cases}$ where $k > 1$ is an integer.

(c, d*) Solve analogues of (a, b) where f is not a polynomial but a function $f : \mathbb{R}^2 \to \mathbb{R}$. (The function f is not assumed to be continuous.)

The system of equations from 8.3.28 (a) is satisfied, for example, by the polynomials

$$f(x, y) = x - y, \quad p(u, v) = u^2 - 4v, \quad \text{and} \quad q(u, v, w) = \frac{u + w}{2}.$$

Below, we assume that $f, g \in \mathbb{R}[x, y]$.

8.3.29. (a) **Lemma.** If $fg = 0$, then $f = 0$ or $g = 0$.

Warning. There exist functions $F, G : \mathbb{R} \to \mathbb{R}$ such that $FG = 0$, $F \neq 0$, and $G \neq 0$. Furthermore, there exist two distinct polynomials (in two variables) which are equal at an infinite set of points. Do not use without proof the fact that if the values of polynomials in two variables are equal at each point, then the polynomials are equal.

(b) If $f^2 = g^2$, then $f = g$ or $f = -g$.

(c) If $f^2 + fg + g^2 = 0$, then $f = 0$ or $g = 0$.

(d) If $f^3 = g^3$, then $f = g$.

(e) If $f^5 = g^5$, then $f = g$.

(f) $f^5 - g^5 = (f - g)(f - \varepsilon_5 g)(f - \varepsilon_5^2 g)(f - \varepsilon_5^3 g)(f - \varepsilon_5^4 g)$.

To prove the assertions 8.3.28 (b, d), the following notions and lemma are useful.

A polynomial f in two variables x and y is called *symmetric* if $f(x, y) = f(y, x)$ and *antisymmetric* if $f(x, y) = -f(y, x)$.

8.3.30. (a) **Lemma.** If $f \in \mathbb{R}[x, y]$ is a polynomial with real coefficients in two variables such that f^2 is symmetric, then f is either symmetric or antisymmetric.

(b) **Lemma.** If $f \in \mathbb{R}[x, y]$ is such that f^{2k+1} is symmetric, then f is symmetric.

(c) If $f \in \mathbb{R}[x, y]$ is antisymmetric, then there exists a symmetric polynomial $a \in \mathbb{R}[x, y]$ such that $f = (x - y)a$.

Lemma 8.3.29 (a) is useful for proving the above assertions, as well as other problems.

8.3.31. Which of the statements in 8.3.29 and 8.3.30 can be generalized to polynomials with complex coefficients?

Next, we develop a generalization of assertion 8.3.28, for an arbitrary number of steps in the definition of the expressibility in radicals.

8.3.32. A *rational function* is a "formal ratio of polynomials," i.e., a pair $f/g := (f, g)$ of polynomials with $g \neq 0$, subject to the equivalence $f/g \sim f'/g'$ when $fg' = f'g$. The polynomial f is identified with the pair $(f, 1)$. Denote by $\mathbb{R}(u_1, \ldots, u_n)$ the set of all rational functions with real coefficients in variables $u_1, \ldots, u_n$.

(a) Define the sum and the product of rational functions. Are they well-defined? Check this!

(b) Consider the system of equations described in Remark 8.2.1 (d) for $n = 2$, where f_j and p_j are rational functions (not necessarily polynomials). Assume that the system is *minimal*. This means that there exists no system with a smaller s, and that f_j^k is not a rational function of $x + y$, xy, and $f_1, \ldots, f_{j-1}$ for any $j = 1, \ldots, s$ and $k < k_j$. Then $s = 1$, $k_1 = 2$, and there exists a rational function $a \in \mathbb{R}(u, v)$ such that

$$f_1(x, y) = (x - y)a(x + y, xy).$$

(c)* State and prove the analogue of (a) where the rational functions $f_1, \ldots,$
f_s are replaced by functions $\mathbb{R}^2 \to \mathbb{R}$ (although $p_0, \ldots, p_s$ are still rational functions) and the equalities for rational functions are replaced by equalities for functions defined on $\mathbb{R}^2$.

8.3.33. (Challenge.) There is only one way to solve a cubic equation. (To solve this problem subsections 3.A and 3.C would be useful.)

Suggestions, solutions, and answers

8.3.28. (a) We will prove that there exists $\alpha \in \mathbb{R}$ such that $f(x, y) = \alpha(x - y)$. Since the polynomial $f^2 = p$ is symmetric, we can assume that the polynomial q is linear in the third variable, i.e., $q(u, v, w) = a(u, v) + b(u, v)w$ for some $a, b \in \mathbb{R}[u, v]$ (otherwise we can change q while preserving f and p). Then we have $x = a(x + y, xy) + b(x + y, xy)f(x, y)$.

This yields $pb^2 = f^2b^2 = (x - a)^2 = (y - a)^2$. By Lemma 8.3.29 (b), we have $x - a = a - y$, since $x - a = y - a$ is impossible. Hence $a = (x + y)/2$, which implies that $(x - y)^2 = 4f^2b^2 = 4pb^2$. If the polynomial $p = f^2$ is constant, the polynomial $b = \pm(x - y)/2\sqrt{p}$ is not symmetric. Therefore p

is not constant. Thus b is constant. Hence $2x = 2q = x + y + 2bf$, so $b \neq 0$ and $f = \alpha(x - y)$ for $\alpha = 1/2b$.

(b) We will prove that k is even and that there exists $\alpha \in \mathbb{R}$ such that $f(x, y) = \alpha(x - y)$. We can use induction on k with the application of part (a) and the generalization of Lemmas 8.3.29 (b, e) and 8.3.30. If k is odd, from Lemma 8.3.30 (b) we get that f is symmetric. This contradicts the equality $x = q(x + y, xy, f(x, y))$. If $k = 4$, then f^2 is either symmetric or antisymmetric. The first case reduces to (a). The second leads to $f^2(x, y) + f^2(y, x) = 0$. The even-$k$ case is similar.

(c) Similarly to part (a) we get $x = a + bf$. Therefore, f is a rational function. The rest of the solution is analogous to (a).

8.3.29. (a) Define the *leading term* of a polynomial so that the leading term of the product is equal to the product of the leading terms of the factors.

(b) This follows from part (a).

(c) We have $f^2 + fg + g^2 = \left(f + \frac{g}{2}\right)^2 + \frac{3}{4}g^2 = (f - \varepsilon_3 g)(f - \varepsilon_3^2 g)$.

(d) This follows from part (c).

(e) This follows from part (f).

(f) Prove and apply Bezout's Theorem for polynomials in u with coefficients in $\mathbb{R}[v]$.

8.3.30. (a) Since f^2 is symmetric, we have $f(x, y)^2 = f(y, x)^2$. By 8.3.29 (b), we have $f(x, y) = \pm f(y, x)$.

(b) See 8.3.29 (c, e).

(c) See the hint for 8.3.29 (f).

8.3.31. *Answer:* 8.3.29 (a, b, f), 8.3.30 (a, b, c).

3.F. Insolvability "in real polynomials" (2)

In this subsection we often omit the arguments (x, y, z) of polynomials.

8.3.34. There are no polynomials $f(x, y, z)$, $p(u, v, w)$, and $q(u, v, w, \tau)$ with real coefficients such that
$$\begin{cases} f(x, y, z)^k = p(\sigma_1(x, y, z), \sigma_2(x, y, z), \sigma_3(x, y, z)), \\ x = q(\sigma_1(x, y, z), \sigma_2(x, y, z), \sigma_3(x, y, z), f(x, y, z)) \end{cases}$$

(a) for $k = 1$;   (b) for $k = 3$;   (c) for $k = 2$;
(d) for any integer $k > 0$.

For the proof, the following definition and statement are useful. A polynomial $f \in \mathbb{R}[x, y, z]$ is called *cyclically symmetric* if $f(x, y, z) = f(y, z, x)$.

8.3.35. If $f \in \mathbb{R}[x, y, z]$ and the polynomial
(a) f^3; (b) f^2
is cyclically symmetric, then f is cyclically symmetric.

Remark 8.3.36 (Cf. solution of problem 8.2.3 (c)). There are no polynomials

$$f_1(x, y, z), \quad f_2(x, y, z), \quad p_0(u, v, w), \quad p_1(u, v, w, \tau_1), \quad p_2(u, v, w, \tau_1, \tau_2)$$

with real coefficients such that

$$\begin{cases} f_1^2 = p_0(\sigma_1, \sigma_2, \sigma_3), \\ f_2^3 = p_1(\sigma_1, \sigma_2, \sigma_3, f_1), \\ x = p_2(\sigma_1, \sigma_2, \sigma_3, f_1, f_2). \end{cases}$$

A generalization of Remark 8.3.36 to an arbitrary number of steps can be formalized by the definition of *expressibility in real radicals*, which is obtained from its complex analogue (2.B) by replacing complex coefficients with real coefficients.

The formulas at the beginning of 2.A show that x is expressible in real radicals for $n = 2$. The solution of problem 8.2.3 (a, b) shows that both of the polynomials

$$(x - y)(y - z)(z - x) \quad \text{and} \quad x^9 y + y^9 z + z^9 x$$

are expressible in real radicals for $n = 3$.

Theorem 8.3.37. The polynomial x is not expressible in real radicals for $n = 3$.

Theorem 8.3.37 is yet another formalization of the fact that *a root of a general cubic equation is not expressible in real radicals in terms of its coefficients*; see Remark 8.1.7 (e). Theorem 8.3.37 is implied by the following lemma.

Lemma 8.3.38 (Preservation of cyclic symmetry). If $q > 0$ is an integer, $f \in \mathbb{R}[x, y, z]$, and the polynomial f^q is cyclically symmetric, then f is cyclically symmetric.

This lemma (in a more general form) will be proved in 4.B.

8.3.39. Which of the statements in this subsection have true analogues for polynomials with complex coefficients?

Suggestions, solutions, and answers

8.3.37. For $n = 3$, the set of polynomials expressible in real radicals is contained in the set of cyclically symmetric polynomials. This statement can be proved by induction on the number of operations in the definition of expressibility in radicals. The inductive step follows from Lemma 8.3.38 on the preservation of cyclic symmetry. Since the polynomial x is not cyclically symmetric, it is not expressible in real radicals.

8.3.38. The proof can be found in 4.B.

8.3.39. *Answer*: 8.3.34 (a, b, c, d), 8.3.35 (b), 8.3.38 for all q which are not divisible by 3.

3.G. Insolvability "in polynomials" (3)

The definition of expressibility in radicals for a polynomial was given on page 133. Formally, Ruffini's Theorem 8.2.2 follows from Lemma 8.3.43, whose formulation is the most interesting and difficult task. In order to do this, we prove the following simple facts. (Clearly, the polynomial x is not a polynomial of $x + y$ and xy.)

8.3.40. The polynomial x_1 is not expressible in radicals in such a way that the second operation in the definition of expressibility is applied only for
    (a) $k = 2$ (*hint:* see problem 8.3.39);   (b) $k = 3$.

8.3.41. Which of the following assertions are true for $f \in \mathbb{C}[x_1, \ldots, x_5]$?
    (a) If f^3 is cyclically symmetric, then f is cyclically symmetric.
    (b) If f^5 is cyclically symmetric, then f is cyclically symmetric.
    (c) If f^3 is symmetric, then f is symmetric.
    (d) If f^2 is symmetric, then f is symmetric.

A 3-cycle is a permutation of an n-element set which moves 3 elements cyclically and does not change the positions of any other elements. A polynomial $f \in \mathbb{C}[x_1, \ldots, x_n]$ is *even-symmetric* if for any 3-cycle α the polynomials $f(x_1, x_2, \ldots, x_n)$ and $f(x_{\alpha(1)}, x_{\alpha(2)}, \ldots, x_{\alpha(n)})$ are equal.

8.3.42. (a) Find a cyclically symmetric polynomial that is not even-symmetric.

(b) Let us assume that a permutation does not change the polynomial from the solution to problem 8.3.41 (d). Then this permutation can be represented as a composition of 3-cycles.

Lemma 8.3.43 (Preservation of even symmetry). If $q > 0$ is an integer,

$f \in \mathbb{C}[x_1, \ldots, x_5]$, and the polynomial f^q is even-symmetric, then f is even-symmetric.

8.3.44. Suppose $f \in \mathbb{C}[x_1, \ldots, x_n]$ is a polynomial.
  (a) If the polynomial f^7 is even-symmetric, then f is even-symmetric.
  (b) If $n \geq 5$ and the polynomial f^3 is even-symmetric, then f is even-symmetric.
  (c) If $n \geq 5$, then any 3-cycle on an n-element set can be written as a product of permutations of the form $(ab)(cd)$ where a, b, c, and d are pairwise distinct (i.e., as a product of compositions of transpositions with disjoint supports).

Lemma 8.3.43 follows from 8.3.44 (a, b) (and from the obvious generalization of (a)). (b) follows from (c). For details, see 4.C.

8.3.45. The definition of *rational* expressibility in real (complex) radicals is analogous to the definition of expressibility in radicals. Polynomials are replaced by rational functions (with appropriate coefficients; see the definition in problem 8.3.32). Is the polynomial x_1 rationally expressible by
  (a) real radicals for $n = 3$?
  (b) (complex) radicals for $n = 5$?

Suggestions, solutions, and answers

8.3.40. (b) Use the analogue of problem 8.3.41 (c) for $n = 3$.

8.3.41. *Answer:* (c) true; (a, b, d) false.
  (a) See 8.2.4 (a).
  (b) Consider the polynomial $x_1 + \varepsilon_5 x_2 + \varepsilon_5^2 x_3 + \varepsilon_5^3 x_4 + \varepsilon_5^4 x_5$.
  (d) Consider the polynomial $\prod_{i<j}(x_i - x_j)$.
  (c) Since f^3 is symmetric, we have

$$f^3(x_1, x_2, x_3, x_4, x_5) = f^3(x_2, x_1, x_3, x_4, x_5).$$

Taking cube roots yields

$$f(x_1, x_2, x_3, x_4, x_5) = \varepsilon_3^q f(x_2, x_1, x_3, x_4, x_5) = \varepsilon_3^{2q} f(x_1, x_2, x_3, x_4, x_5).$$

Thus $\varepsilon_3^{2q} = 1$, so $\varepsilon_3^q = 1$. Similarly, $f(\vec{x}) = f(\vec{x}_\alpha)$ for any permutation α exchanging two elements from the set $\{x_1, x_2, x_3, x_4, x_5\}$. Therefore f is symmetric.

8.3.42. (a) $x_1 x_2 + x_2 x_3 + x_3 x_4 + x_4 x_5 + x_5 x_1$.
  (b) First prove that if a permutation maps the polynomial of problem 8.3.41 (d) to itself, then the permutation is even. This implies that the

permutation can be represented as a composition of 3-cycles; see Chapter 4, section 2.

8.3.43. See the discussion of Ruffini's Theorem 8.4.4.

8.3.44. (c) Let a, b, c, d, and e be five distinct elements of the given set. Then $(abc) = (ac)(de)(ab)(de)$.

8.3.45. *Answer:* (a, b) No.

Lemmas 8.3.38 on preserving cyclic symmetry and 8.3.43 on preserving even symmetry also hold for rational functions; see 8.4.4. After that, we use the ideas in the solution to 8.3.37.

3.H. Insolvability in complex numbers (4*)

8.3.46. (a) Let $x, y, r \in \mathbb{R}$, $p, g \in \mathbb{Q}[u, v]$, and $p_1 \in \mathbb{Q}[u, v, w]$ be such that $g(x, y) \notin \mathbb{Q}(x + y, xy)$ and

$$\begin{cases} r^2 = p(x + y, xy), \\ g(x, y) = p_1(x + y, xy, r) \end{cases}$$

(cf. problem 8.3.28 (c)). Then $r \in \mathbb{Q}(x, y)$.

(b) Let $x, y, r \in \mathbb{R}$, $p \in \mathbb{Q}[\sqrt{2}][u, v]$, $g \in \mathbb{Q}[u, v]$, and $p_1 \in \mathbb{Q}[\sqrt{2}][u, v, w]$ be such that $g(x, y) \notin \mathbb{Q}(x + y, xy, \sqrt{2})$ and the equations of (a) hold. Then there exist $\rho \in \mathbb{Q}(x, y)$, $\pi \in \mathbb{Q}[\sqrt{2}][u, v]$, and $\pi_1 \in \mathbb{Q}[\sqrt{2}][u, v, w]$ such that the equations of (a) hold with r, p, and p_1 replaced by ρ, π, and π_1.

(c) **Rationalization lemma.** Let $x, y, r \in \mathbb{R}$ and let $F \subset \mathbb{R}$ be a field containing $x + y, xy, r^2$ but not r. If $F(r) \cap \mathbb{Q}(x, y) \not\subset F$, then there exists $\rho \in \mathbb{Q}(x, y)$ such that $\rho^2 \in F$ and $F(\rho) = F(r)$.

8.3.47. Let $a_j = \sigma_j(x_1, x_2, x_3)$, $j = 1, 2, 3$.

(a) Let $x_1, x_2, x_3, r \in \mathbb{R}$, $p, g \in \mathbb{Q}[u_1, u_2, u_3]$ and $p_1 \in \mathbb{Q}[u_1, u_2, u_3, v]$ be such that $g(x_1, x_2, x_3) \notin \mathbb{Q}(a_1, a_2, a_3)$ and

$$\begin{cases} r^2 = p(a_1, a_2, a_3), \\ g(x_1, x_2, x_3) = p_1(a_1, a_2, a_3, r). \end{cases}$$

Then $r \in \mathbb{Q}(x_1, x_2, x_3)$.

(b) **Rationalization lemma.** Let $x_1, x_2, x_3, r \in \mathbb{R}$ and let $F \subset \mathbb{R}$ be a field containing a_1, a_2, and a_3, r^2 but not r. If $F(r) \cap \mathbb{Q}(x_1, x_2, x_3) \not\subset F$, then there exists $\rho \in \mathbb{Q}(x_1, x_2, x_3)$ such that $\rho^2 \in F$ and $F(\rho) = F(r)$.

(c) **Proposition.** If $x_1, x_2, x_3 \in \mathbb{R}$ and x_1 is $\{a_1, a_2, a_3\}$-expressible by quadratic real radicals, then x_1 is $\{a_1, a_2, a_3\}$-expressible by quadratic real radicals so that every radical is in $\mathbb{Q}(x_1, x_2, x_3)$.

8.3.48. (a) Let $x, y, r \in \mathbb{C}$, $p \in \mathbb{Q}[u, v]$, and $p_1 \in \mathbb{Q}[u, v, w]$ be such that

$$\begin{cases} r^3 = p(x + y, xy), \\ x = p_1(x + y, xy, r) \end{cases}$$

(cf. problem 8.3.28 (d) for $k = 3$). Then $r \in \mathbb{Q}[\varepsilon_3](x, y)$.

(b) Same as (a), but with $x = p_1(x + y, xy, r)$ replaced by $g(x, y) = p_1(x + y, xy, r)$ for some $g \in \mathbb{Q}[u, v]$ such that $g(x, y) \notin \mathbb{Q}(x + y, xy)$.

(c) **Rationalization lemma.** Let $x, y, r \in \mathbb{C}$ and let $F \subset \mathbb{C}$ be a field containing $x + y$, xy, ε_3, and r^3 but not r. If $F(r) \cap \mathbb{Q}(x, y) \not\subset F$, then there exists $\rho \in \mathbb{Q}(x, y)$ such that $\rho^3 \in F$ and $F(\rho) = F(r)$.

(d) **Rationalization lemma.** Same as (c) with x and y replaced by x_1, $\ldots, x_n$ and with $x + y$ and xy replaced by $\sigma_1(x_1, \ldots, x_n), \ldots, \sigma_n(x_1, \ldots, x_n)$.

(e) **Rationalization lemma.** Same as (d) with r^3 and ρ^3 replaced by r^q and ρ^q for a prime q and with ε_3 replaced by ε_q.

(f) **Proposition.** If

$$x_1, \ldots, x_n \in \mathbb{C}, \quad M := \{\sigma_1(x_1, \ldots, x_n), \ldots, \sigma_n(x_1, \ldots, x_n),\}$$

and x_1 is M-expressible in radicals, then x_1 is M-expressible in radicals so that every radical is in $\bigcup_{q=3}^{\infty} \mathbb{Q}[\varepsilon_q](x_1, \ldots, x_n)$.

8.3.49. There exist numbers $x, y \in \mathbb{R}$ such that if $p \in \mathbb{Q}[u, v]$ and $p(x, y) = 0$, then $p = 0$.

Such numbers are called *algebraically independent over* $\mathbb{Q}$.

3.I. Expressibility with a given number of radicals (4*)

Definitions of the expressibility in radicals for a number and a polynomial are given in 1.D and 2.A, respectively.

8.3.50. The roots of cubic and quartic equations with rational coefficients are expressible in radicals with root extraction made only

(a) twice, with one cube root and one square root for cubic equations;

(b) four times, with one cube root and three square roots for quartic equations.

("Once" means "used once in an algorithm." For example, in the algorithm $u := \sqrt[3]{a}$, $v := u + u$, the cube root is extracted once.)

8.3.51. The roots of a cubic equation (as polynomials) are not expressible in radicals with extraction of only
   (a) one root;
   (b) square roots;
   (c) cube roots;
   (d)* "single-level" roots, i.e., roots of non-radical expressions.

8.3.52. Are the roots of a 4th-degree equation (as polynomials) expressible in radicals with extraction of only
   (a) square roots;   (b) cube roots;
   (c) two roots;   (d) three roots?

8.3.53. If the roots of an equation of nth degree with rational coefficients are radical, then they are radical with extraction of no more than
   (a) five roots for $n = 5$;   (b) $n \log_2 n$ roots for any n.

   To solve these problems we need the following simple elements of Galois theory. Define $\mathbb{Z}_q := \{1, \varepsilon_q, \varepsilon_q^2, \ldots, \varepsilon_q^{q-1}\}$. Recall that S_n is the set of all permutations of a set of n elements. A subset of S_n is called a *subgroup* if it is closed under composition and taking inverses. For a subgroup $G \subset S_n$ and an integer q, a map $G \to \mathbb{Z}_q$ is called a *homomorphism* (or *character*) if it maps compositions into products, i.e., if $\chi(\alpha\beta) = \chi(\alpha)\chi(\beta)$.

8.3.54. (a) If $q > 0$ is an integer, f is a nonzero polynomial, and f^q is even-symmetric, then for any even permutation α there exists a unique character $\chi_f(\alpha) \in \mathbb{Z}_q$ such that $f(x_{\alpha(1)}, x_{\alpha(2)}, \ldots, x_{\alpha(n)}) = \varepsilon_q^{\chi_f(\alpha)} f(x_1, x_2, \ldots, x_n)$.
   (b) The map $\chi_f : A_n \to \mathbb{Z}_q$ from the set A_n of all even permutations constructed in (a) is a homomorphism.

8.3.55. Does there exist a prime q and a non-constant homomorphism
   (a) $A_3 \to \mathbb{Z}_q$?   (b) $A_4 \to \mathbb{Z}_q$?

   In the proof of Lemma 8.3.43 on the preservation of even symmetry, it was actually proved that for an integer $n \geq 5$ and a prime q, every homomorphism $\chi : A_n \to \mathbb{Z}_q$ must map each permutation to 1.

8.3.56. (a) Do there exist an integer q and an injective (one-to-one) homomorphism $S_3 \to \mathbb{Z}_q$?
   (b) Do there exist integers p and q and homomorphisms $\chi : S_4 \to \mathbb{Z}_q$ and $\varphi : \chi^{-1}(1) \to \mathbb{Z}_p$ with the second one being injective?

8.3.57. (a) Do there exist integers p and q and homomorphisms $\chi : S_4 \to \mathbb{Z}_q$ and $\varphi : \chi^{-1}(1) \to \mathbb{Z}_p$ with the second one being injective?

(b) Do there exist integers p, q, and r and homomorphisms $\chi : S_4 \to \mathbb{Z}_q$, $\varphi : \chi^{-1}(1) \to \mathbb{Z}_p$, and $\gamma : \varphi^{-1}(1) \to \mathbb{Z}_r$ with the last one being injective?

(c) Does there exist a chain of four homomorphisms analogous to (b)?

8.3.58. (a) For any polynomial $f(x_1, x_2, \ldots, x_n)$, the set
$$st_f := \{\alpha \in S_n \mid f(x_{\alpha(1)}, x_{\alpha(2)}, \ldots, x_{\alpha(n)}) = f(x_1, x_2, \ldots, x_n)\}$$
is a subgroup of S_n.

(b) List all subgroups of S_3. Which of them can be preimages of the identity element under the homomorphism $S_3 \to \mathbb{Z}_q$ for some q?

(c) List all subgroups in S_4. Which of them can be preimages of the identity element under the homomorphism $S_4 \to \mathbb{Z}_q$ for some q?

The estimation 8.3.53 can be obtained from the arguments in the proof of Ruffini's Theorem 8.2.2. The idea is that the "symmetry subgroup" st_f of S_n cannot be changed more than $\log_2(n!) < n \log_2 n$ times.

4. Proofs of insolvability in radicals

Formally, to understand this section, it is enough to read subsections 1.B–1.D (although we sometimes refer to section 3 for some details of the proofs below, and we use Gauss's degree-lowering theorem (8.1.15) in subsection 4.G below). The beginning of section 3 lists which statements are helpful to understand this section.

Guide to this section. Subsection 4.A is used throughout. Otherwise, the subsections are formally independent of one another with the following exceptions:

Subsection 4.G uses Lemma 8.4.14. Theorem 8.4.4 is used in subsection 4.F. Subsection 4.G is needed for subsection 4.H.

Nevertheless, it is useful to read the subsections in the following order:

4.B before 4.C, 4.D before 4.E before either 4.F or 4.G.

The outline of the ideas are provided by statements of the lemmas in each subsection.

4.A. Fields and their extensions (2)

If $F \subset \mathbb{C}$, $r \in \mathbb{C}$, and $r^q \in F$ for some positive integer q, then define
$$F[r] := \{a_0 + a_1 r + a_2 r^2 + \cdots + a_{q-1} r^{q-1} \mid a_0, \ldots, a_{q-1} \in F\}.$$

Definitions of constructibility, real constructibility, expressibility in radicals, and real expressibility in radicals are given in 2.C, 1.B, 1.D, and 1.C, respectively.

Lemma 8.4.1 (Tower of extensions). (a) A number $x \in \mathbb{C}$ is constructible if and only if there exist $r_1, \ldots, r_{s-1} \in \mathbb{C}$ such that

$$\mathbb{Q} = F_1 \subset F_2 \subset F_3 \subset \ldots \subset F_{s-1} \subset F_s \ni x,$$

where $r_k^2 \in F_k$, $r_k \notin F_k$, and $F_{k+1} = F_k[r_k]$ for every $k = 1, \ldots, s-1$.

(b) A number $x \in \mathbb{C}$ is expressible in radicals if and only if there exist $r_1, \ldots, r_{s-1} \in \mathbb{C}$ and primes $q_1, \ldots, q_{s-1}$ such that

$$\mathbb{Q} = F_1 \subset F_2 \subset F_3 \subset \ldots \subset F_{s-1} \subset F_s \ni x,$$

where $r_k^{q_k} \in F_k$, $r_k \notin F_k$, and $F_{k+1} = F_k[r_k]$ for every $k = 1, \ldots, s-1$.

Such a sequence is called a *tower of* (*quadratic* or *radical*) *extensions*.

This lemma is proved by induction on the number of operations required to obtain the given number, similar to previous lemmas in section 3. An analogue of this lemma for real constructibility and radicality is valid and is proved similarly.

The concept of a *field* will help us to think about this natural but somewhat cumbersome lemma. A field is a subset of $\mathbb{C}$ which is closed under addition, subtraction, multiplication, and division by a nonzero number. The conventional name is "number field" (the technical term "field" in mathematics refers to a more general object). This notion is useful because the Polynomial Remainder Theorem holds for polynomials with coefficients in a field. We use the standard notation $F[u_1, \ldots, u_n]$ and $F(u_1, \ldots, u_n)$ for the sets of polynomials and rational functions (i.e., formal ratios of polynomials) with coefficients in a field F. Equality signs involving polynomials or rational functions P, f, or f_j mean coefficientwise equality of polynomials or rational functions.

Recall the notation

$$\varepsilon_q := \cos(2\pi/q) + i\sin(2\pi/q) \quad \text{and} \quad \vec{y} := (y_1, \ldots, y_n).$$

4.B. Insolvability "in real polynomials" (3)

Here we prove Theorem 8.4.2, the real version of Ruffini's Theorem 8.4.4 below. Define *an extension of a field* $F \subset \mathbb{C}$ *by* $r_1, \ldots, r_s \in \mathbb{C}$ as

$$F(r_1, \ldots, r_s) := \{P(r_1, \ldots, r_s) \mid P \in F(u_1, \ldots, u_s)\}.$$

Define a *radical extension* $F[r_1, \ldots, r_s]$ of a field $F \subset \mathbb{C}$ inductively by $F[r_1, \ldots, r_s] := F[r_1, \ldots, r_{s-1}][r_s]$, where we assume that for every $j = 1, \ldots, s$ there is an integer q_j such that $r_j^{q_j} \in F[r_1, \ldots, r_{j-1}]$.

Theorem 8.4.2. There exist $a_0, a_1, a_2 \in \mathbb{R}$ such that the equation $x^3 + a_2 x^2 + a_1 x + a_0 = 0$ has three real roots x_1, x_2, x_3, none of which lies in any radical extension of $\mathbb{Q}(a_0, a_1, a_2)$ contained in $\mathbb{Q}(x_1, x_2, x_3)$.

A rational function $P \in \mathbb{R}(u_1, u_2, u_3)$ is said to be *cyclically symmetric* if $P(u_1, u_2, u_3) = P(u_2, u_3, u_1)$ (cf. Remark 8.1.7 (e).)

Lemma 8.4.3 (Preservation of cyclic symmetry). If P is a rational function of three variables with coefficients in $\mathbb{R}$ and P^q is cyclically symmetric for some integer q, then P is cyclically symmetric.

Proof. Let $R(x_1, x_2, x_3) := P(x_2, x_3, x_1)$. Since P^q is cyclically symmetric, we have $P^q = R^q$.

If q is odd, we obtain $P = R$ (similar to problem 8.3.29), so P is cyclically symmetric.

Otherwise, if q is even, then $P = R$ or $P = -R$. When $P = R$ we obtain that P is cyclically symmetric. When $P = -R$ we have

$$P(x_1, x_2, x_3) = -P(x_2, x_3, x_1) = P(x_3, x_1, x_2) = -P(x_1, x_2, x_3).$$

Thus $P = 0$, so P is cyclically symmetric again. □

Proof of Theorem 8.4.2. The numbers $x_1, \ldots, x_n \in \mathbb{R}$ are called *algebraically independent* over $\mathbb{Q}$ if $P(x_1, \ldots, x_n) \neq 0$ for every nonzero polynomial P with coefficients in $\mathbb{Q}$. By induction we show that *for any n there are n algebraically independent numbers $x_1, \ldots, x_n$ over $\mathbb{Q}$.* The inductive step follows because $\mathbb{R}$ is uncountable, whereas the set of real roots of polynomials with coefficients in $\mathbb{Q}(x_1, \ldots, x_{n-1})$ is countable.

Denote the coefficients of the monic polynomial with roots x_1, x_2, x_3 by

$$a_2 := -(x_1 + x_2 + x_3), \quad a_1 = x_1 x_2 + x_2 x_3 + x_1 x_3, \quad a_0 = -x_1 x_2 x_3.$$

Assume to the contrary that there is a radical extension $\mathbb{Q}(a_0, a_1, a_2)[r_1, \ldots, r_s]$ which both contains x_1 and is contained in $\mathbb{Q}(x_1, x_2, x_3)$. Using Lemma 8.4.3, by induction on j we see that r_j is the value at (x_1, x_2, x_3) of an even-symmetric rational function for every $j = 1, \ldots, s$. As $x_1 \in \mathbb{Q}(a_0, a_1, a_2)[r_1, \ldots, r_s]$, we see that x_1 is also the value at (x_1, x_2, x_3) of an even-symmetric rational function P_0. Since x_1, x_2, x_3 are algebraically independent over $\mathbb{Q}$, we have $P_0(u_1, u_2, u_3) = u_1$. This is not even-symmetric because $P_0(u_2, u_3, u_1) = u_2 \neq u_1$, which is a contradiction. □

4.C. Insolvability "in polynomials" (3)

Here we prove Ruffini's Theorem in the following form, used by Theorem 8.1.13 (cf. Ruffini's Theorem 8.2.2). Extensions and radical extensions were defined at the beginning of 4.B. Define

$$\mathbb{Q}_\varepsilon := \bigcup_{q=3}^{\infty} \mathbb{Q}(\varepsilon_3, \varepsilon_4, \ldots, \varepsilon_q)$$

and

$$\mathbb{Q}_\varepsilon(\vec{a}) := \mathbb{Q}_\varepsilon(a_0, a_1, a_2, a_3, a_4).$$

Theorem 8.4.4 (Ruffini). There exist $a_0, a_1, a_2, a_3, a_4 \in \mathbb{C}$ such that no root of the equation $x^5 + a_4 x^4 + \ldots + a_1 x + a_0 = 0$ is contained in any radical extension of $\mathbb{Q}_\varepsilon(\vec{a})$ contained in $\mathbb{Q}_\varepsilon(\vec{x})$, where $x_1, \ldots, x_5$ are the roots of the equation.

In order to understand the main idea, one can replace $\mathbb{Q}_\varepsilon$ by $\mathbb{Q}$ and $\mathbb{Q}_\varepsilon(\vec{x})$ by $\mathbb{Q}[\vec{x}]$ (in the statement and proof).

For a permutation α write

$$\vec{u}_\alpha := (u_{\alpha(1)}, \ldots, u_{\alpha(n)}).$$

A rational function $P \in \mathbb{C}(\vec{u})$ is *even-symmetric* if $P(\vec{u}) = P(\vec{u}_{(abc)})$ for every cycle (abc) of length three.[10]

Lemma 8.4.5 (Preserving symmetry after root extractions). *If P is a rational function of five variables with coefficients in $\mathbb{C}$ and P^q is even-symmetric for some integer q, then P is even-symmetric.*

Proof. We may assume that q is a prime and $P \neq 0$.

Let $\{a, b, c, d, e\} = \{1, \ldots, 5\}$.

First assume that $q \neq 3$. Since $P^q(\vec{u}) = P^q(\vec{u}_{(abc)})$, we have

$$\prod_{j=0}^{q-1} (P(\vec{u}) - \varepsilon_q^j P(\vec{u}_{(abc)})) = 0.$$

Since P is a nonzero rational function, there exists $j = j(abc) \in \mathbb{Z}$ such that

$$P(\vec{u}) = \varepsilon_q^j P(\vec{u}_{(abc)}).$$

Then

$$P(\vec{u}) = \varepsilon_q^j P(\vec{u}_{(abc)}) = \varepsilon_q^{2j} P(\vec{u}_{(abc)^2}) = \varepsilon_q^{3j} P(\vec{u}).$$

Hence j is divisible by q, i.e., $P(\vec{u}) = P(\vec{u}_{(abc)})$.

For $q = 3$ let $\sigma := (ab)(de) = (abe)(bed)$. Then analogously $0 \equiv j(\sigma^2) \equiv 2j(\sigma) \pmod{3}$. Hence $j(\sigma) \equiv 0 \pmod{3}$; i.e., $P(\vec{u}) = P(\vec{u}_\sigma)$. Analogously, $P(\vec{u}) = P(\vec{u}_{(ac)(de)})$. Since $(ab)(de)(ac)(de) = (abc)$, we have $P(\vec{u}) = P(\vec{u}_{(abc)})$. $\square$

Proof of Ruffini's Theorem 8.4.4. The quantities $x_1, \ldots, x_n \in \mathbb{C}$ are said to be *algebraically independent* over $\mathbb{Q}_\varepsilon$ if $P(\vec{x}) \neq 0$ for every nonzero polynomial P with coefficients in $\mathbb{Q}_\varepsilon$. By induction on n *there are n algebraically independent numbers $x_1, \ldots, x_n$ over $\mathbb{Q}_\varepsilon$.* The inductive step follows because $\mathbb{C}$ is uncountable while the set of roots of polynomials with coefficients in $\mathbb{Q}_\varepsilon(x_1, \ldots, x_{n-1})$ is countable.

Denote the coefficients of the monic polynomial with roots $x_1, \ldots, x_5$ by

$$a_4 := -(x_1 + \ldots + x_5), \quad \ldots, \quad a_0 = -x_1 \cdot \ldots \cdot x_5.$$

---

[10]A permutation is *even* if it is a composition of an even number of transpositions. Being even-symmetric is equivalent to $P(\vec{u}) = P(\vec{u}_\alpha)$ for every even permutation α of $\{1, \ldots, n\}$. Indeed, any even permutation is a composition of permutations of the form $(ab)(bc) = (abc)$ and $(ab)(cd) = (abc)(bcd)$.

Assume to the contrary that there is a radical extension $\mathbb{Q}_\varepsilon(\vec{a})[r_1, \ldots, r_s]$ which both contains x_1 and is contained in $\mathbb{Q}_\varepsilon(\vec{x})$. Using Lemma 8.4.5, by induction on j we see that r_j is the value at $\vec{x}$ of an even-symmetric rational function for every $j = 1, \ldots, s$. Since $x_1 \in \mathbb{Q}_\varepsilon(\vec{a})[r_1, \ldots, r_s]$, we see that x_1 is also the value at $\vec{x}$ of an even-symmetric rational function P_0. Since $x_1, \ldots, x_5$ are algebraically independent over $\mathbb{Q}_\varepsilon$, we have $P_0(u_1, \ldots, u_5) = u_1$. This is not even-symmetric because the cycle (123) carries u_1 to $u_2 \neq u_1$, so we have a contradiction. $\qquad\square$

4.D. Non-constructibility in Gauss's Theorem (3*)

Theorem 8.1.2 (asserting the non-constructibility of $\sqrt[3]{2}$) follows from the real analogue of the tower of extensions lemma 8.4.1 (a) and the following result.

Lemma 8.4.6. Let $F \subset \mathbb{R}$ be a field and let $r \in \mathbb{R} - F$ such that $r^2 \in F$.
   (a) Then $F[r]$ is a field.
   (b) If $\sqrt[3]{2} \notin F$, then $\sqrt[3]{2} \notin F[r]$.

Proof. (a) It is necessary to prove that $F[r]$ is closed under addition, subtraction, multiplication, and division by a nonzero number. This is trivial for all operations except division, for which the statement holds because

$$\frac{1}{a+br} = \frac{a}{a^2 - b^2 r^2} - \frac{b}{a^2 - b^2 r^2} r.$$

(b) Suppose to the contrary that $\sqrt[3]{2} \in F[r]$. Then $\sqrt[3]{2} = a + br$ for some $a, b \in F$. We get

$$2 = (\sqrt[3]{2})^3 = (a^3 + 3ab^2 r^2) + (3a^2 b + b^3 r^2)r.$$

Since $\sqrt[3]{2} \notin F$, we have $b \neq 0$ and $r \notin F$. In particular, $r \neq 0$. Therefore $3a^2 + b^2 r^2 > 0$. Since $2 \in \mathbb{Q} \subset F$ we have $r \in F$, a contradiction. $\qquad\square$

Now we prove Gauss's non-constructibility result in 8.1.5.

Lemma 8.4.7 (Powers of 2). If a polynomial with rational coefficients is irreducible over $\mathbb{Q}$ and has a constructible root, then the degree of the polynomial is a power of 2.

This lemma is implied by the tower of extensions lemma 8.4.1 (a) and part (b) of the following lemma. The proof of (a) is left to the reader as an exercise.

Lemma 8.4.8 (Conjugation). Let $F \subset \mathbb{C}$ be a field and let $r \in \mathbb{C} - F$ such that $r^2 \in F$.
   (a) Define the conjugation map $F[r] \to F[r]$ by the formula $\overline{x + yr} := x - yr$. This map is well-defined and we have $\overline{z + w} = \overline{z} + \overline{w}$ and $\overline{zw} = \overline{z} \cdot \overline{w}$.

(b) If polynomials $P \in F[x]$ and $Q \in F[r][x]$ have a common root and are irreducible over F and over $F[r]$, respectively, then $\deg P \in \{\deg Q, 2 \deg Q\}$.

Proof. (b) By the complex analogue of Lemma 8.4.6 (a), $F[r]$ is a field. Consider divisibility, irreducibility, and GCDs in $F[r]$, unless otherwise stated. Since P and Q have a common root and Q is irreducible, P is divisible by Q. By (a) $P = \overline{P}$ is divisible by $\overline{Q}$. Since Q is irreducible and divisible by $D := \gcd(Q, \overline{Q})$, it follows that either $D = Q$ or $D = 1$.

If $D = Q$, then from $\overline{D} = D$ we obtain $Q = D \in F[x]$. Since P is irreducible over F, we obtain $P = Q$.

If $D = 1$, then P is divisible by $M := Q\overline{Q}$. Since $\overline{M} = M$, we have $M \in F[x]$. Since P is irreducible over F, we obtain $P = M$. Hence $\deg P = 2 \deg Q$. $\qquad\square$

Lemma 8.4.9 (Eisenstein's criterion). Let p be a prime. If the leading coefficient of a polynomial with integer coefficients is not divisible by p, the other coefficients are divisible by p, and the constant term is not divisible by p^2, then this polynomial is irreducible over $\mathbb{Z}$.

Lemma 8.4.10 (Gauss's lemma). If a polynomial with integer coefficients is irreducible over $\mathbb{Z}$, then it is irreducible over $\mathbb{Q}$.

Both Eisenstein's criterion and Gauss's lemma are easily proved by passing to polynomials with coefficients in $\mathbb{Z}_p$. (For Gauss's lemma, consider the factorization $P = P_1 P_2$ of the given polynomial P over $\mathbb{Q}$, take n_1 and n_2 such that the polynomials $n_1 P_1$ and $n_2 P_2$ both have integer coefficients, and consider a prime divisor p of $n_1 n_2$. For the Eisenstein criterion, see the solution to problem 8.3.1 (f).)

Proof of non-constructibility in Gauss's Theorem 8.1.5. Since $\varepsilon_n = \varepsilon_{nk}^k$, the constructibility of ε_{nk} implies the constructibility of ε_n. Hence it suffices to prove that ε_n is not constructible for

(a) n being a prime not of the form $2^m + 1$, and

(b) $n = p^2$, the square of a prime.

The non-constructibility of ε_n follows by Lemma 8.4.7 (on powers of 2) for the root ε_n of the polynomial

- $P(x) := x^{n-1} + x^{n-2} + \cdots + x + 1$ for case (a) and
- $P(x) := x^{p(p-1)} + x^{p(p-2)} + \cdots + x^p + 1$ for case (b).

The irreducibility of these polynomials over $\mathbb{Q}$ follows from their irreducibility over $\mathbb{Z}$ and Gauss's lemma 8.4.10. The irreducibility of these polynomials $P(x)$ over $\mathbb{Z}$ follows from irreducibility of $P(x + 1)$ over $\mathbb{Z}$. The latter is a consequence of Eisenstein's criterion, using the congruence $(a + b)^p \equiv a^p + b^p \bmod p$. $\qquad\square$

4.E. Insolvability "in real numbers"

The implication (ii)⇒(i) in Theorem 8.1.8 follows from the real analogue of the tower of extensions lemma 8.4.1 (b) and part (a) of the following.

Lemma 8.4.11. Let q be a prime, $F \subset \mathbb{R}$ a field and $r \in \mathbb{R} - F$ with $r^q \in F$.

(a) If a polynomial with coefficients in F is of degree 3 and has three real roots, none of which lies in F, then none of the roots lies in $F[r]$.

(b) **Irreducibility.** The polynomial $t^q - r^q$ is irreducible over $F[\varepsilon_q]$.

(c) **Linear independence.** If $P(r) = 0$ for some polynomial $P \in F[\varepsilon_q][t]$ of degree less than q, then $P = 0$.

(d) **Conjugation.** If $P \in F[\varepsilon_q][t]$ and $P(r) = 0$, then $P(r\varepsilon_q^k) = 0$ for every $k = 0, 1, \ldots, q - 1$.

Proof of (b). Suppose to the contrary that the polynomial $t^q - r^q$ factors over $F[\varepsilon_q]$, that is, has a proper divisor $P \in F[\varepsilon_q][t]$. The roots of $t^q - r^q$ are $r, r\varepsilon_q, r\varepsilon_q^2, \ldots, r\varepsilon_q^{q-1}$. The constant term of P is the product of k of these roots. Then $r^k \in F[\varepsilon_q]$. Since q is prime, $kx + qy = 1$ for some integers x and y. Then $r = (r^k)^x (r^q)^y \in F[\varepsilon_q]$.

Therefore[11] $r^2, r^3, \ldots, r^{q-1} \in F[\varepsilon_q]$. Consider a $q \times (q-1)$ matrix with entries $a_{kl} \in F$ formed by representations of numbers r^k in powers of ε_q:

$$r^k = \sum_{l=0}^{q-2} a_{kl} \varepsilon_q^l, \qquad 0 \le k \le q - 1.$$

Using additions and multiplications by numbers in F, we can obtain a matrix with a zero row.

Hence there is a nonzero polynomial $Q \in F[t]$ of degree less than q with the root r. Then $\gcd(t^q - r^q, Q)$ has a root r and degree k with $0 < k \le \deg Q < q$. So the polynomial $t^q - r^q$ is reducible over F.

Therefore we see that $r \in F$, a contradiction.   □

Part (c) is similar to (b).

Proof of (c). Since $P(r) = 0$, the remainder on division of P by $t^q - r^q$ assumes value 0 at r.

Since the degree of this remainder is less than q, by (c) this remainder is zero. Thus P is divisible by $t^q - r^q$. For every $j = 0, 1, \ldots, q - 1$, since $(r\varepsilon_q^j)^k = r^k$, we obtain $P(r\varepsilon_q^j) = 0$.   □

Proof of (a). Suppose to the contrary that a root x_0 of the polynomial A lies in $F[r]$. Since x_0 is in $F[r]$ and $r^q \in F$, $x_0 = H(r)$ for some polynomial H with coefficients in F of degree greater than 0 and less than q. Apply (d) to $P(t) := A(H(t))$. Since $A(H(r)) = 0$, we see that $H(r\varepsilon_q^k)$ is the root of A for every $k = 0, 1, \ldots, q - 1$. If $H(r\varepsilon_q^k) = H(r\varepsilon_q^l)$ for some k and l with

---

[11] Using the fact that dimension is well-defined, the above can be rewritten as $\dim_F F[r] \le \dim_F F[\varepsilon_q] \le q - 1$.

$0 \leq k < l \leq q - 1$, then by (c) we have $\deg H = 0$, which is a contradiction. Thus the numbers $H(r\varepsilon_q^k)$, $0 \leq k \leq q - 1$, are distinct roots of A. So $q \leq 3$.

If $q = 2$, then by Vieta's Theorem (3.6.5), the remaining root of A lies in F, a contradiction. Hence $q = 3$. Since $\overline{\varepsilon_3} = \varepsilon_3^2$, we have $\overline{H(r\varepsilon_3)} = H(r\varepsilon_3^2)$. Since the last two numbers are distinct, neither of them is real.   □

4.F. Insolvability "in numbers" (4*)

Theorem 8.1.13 follows from the Ruffini Theorem 8.4.4 and the Abel–Ruffini Theorem 8.4.12.

Theorem 8.4.12 (Abel–Ruffini). Let $a_0, \ldots, a_{n-1} \in \mathbb{C}$ be such that the equation $x^n + a_{n-1}x^{n-1} + \ldots + a_1 x + a_0 = 0$ has n distinct roots $x_1, \ldots, x_n$. If there exists a radical extension $\mathbb{Q}_\varepsilon(\vec{a}) = F_0 \subset \ldots \subset F_s$ such that $x_1 \in F_s$, then there exists a radical extension $\mathbb{Q}_\varepsilon(\vec{a}) = Q_0 \subset \ldots \subset Q_t$ such that $x_1 \in Q_t \subset \mathbb{Q}_\varepsilon(\vec{x})$.[12]

The proof of Theorem 8.4.12 does not use permutations or the assumption that $n \geq 5$. We construct $Q_1, \ldots, Q_t$ inductively using the lemma below, which asserts that if $F[r]$ contains more (values of) rational functions of $x_1, \ldots, x_n$ with coefficients in $\mathbb{Q}_\varepsilon$ than does F, then we may assume that r itself is such an "excess" rational function.

Lemma 8.4.13 (Rationalization). Let $x_1, \ldots, x_5, r \in \mathbb{C}$, let q be a prime, and let $F \subset \mathbb{C}$ be a field containing the elementary symmetric polynomials of $x_1, \ldots, x_5$ and also ε_q and r^q, but not r. If $F[r] \cap \mathbb{Q}_\varepsilon(\vec{x}) \not\subset F$, then there exists $\rho \in \mathbb{Q}_\varepsilon(\vec{x})$ such that $\rho^q \in F$ and $F[\rho] = F[r]$.

Proof. By assumption, there is a rational function $U \in \mathbb{Q}_\varepsilon(\vec{u})$ satisfying $U(\vec{x}) \in F[r] - F$. Hence

$$U(\vec{x}) = P(r) = p_0 + p_1 r + \ldots + p_{q-1} r^{q-1}$$

for some polynomial $P \in F[z]$ of degree less than q. Since $P(r) \notin F$, there exists l such that $0 < l < q$ with nonzero coefficient $p_l \in F$ of z^l in P. We have

$$\rho := p_l r^l = \frac{P(r) + \varepsilon_q^{-l} P(r\varepsilon_q) + \varepsilon_q^{-2l} P(r\varepsilon_q^2) + \cdots + \varepsilon_q^{(1-k)l} P(r\varepsilon_q^{q-1})}{q}.$$

Define the *resolution polynomial* $Q(t) := \prod_{\alpha \in S_5} (t - U(\vec{x}_\alpha))$, where S_5 is the set of all permutations of $\{1, \ldots, 5\}$. Since $U(\vec{x}) = P(r)$, we have $Q(P(r)) = 0$. The coefficients of Q as a polynomial of t are symmetric in $x_1, \ldots, x_5$. Since F contains elementary symmetric polynomials of $x_1, \ldots, x_5$, it follows that $Q(t) \in F[t]$. Thus $Q(P(z)) \in F[z]$. Take any $j = 1, \ldots, q - 1$. Then by

---

[12]Here "$x_1 \in Q_t \subset \mathbb{Q}_\varepsilon(\vec{x})$" can be replaced by "$Q_t = F_s \cap \mathbb{Q}_\varepsilon(\vec{x})$". The proof is analogous.

the conjugation lemma 8.4.14 (c), $Q(P(r\varepsilon_q^j)) = 0$. Thus $P(r\varepsilon_q^j) = U(\vec{x}_\alpha)$ for some permutation $\alpha = \alpha_j$. Since $\varepsilon_q \in F$, the above formula for ρ shows that $\rho \in \mathbb{Q}_\varepsilon(\vec{x})$.

We have $\rho^q = p_l^q(r^q)^l \in F$ and $\rho = p_l r^l \in F[r]$. Since q is a prime and l is not divisible by q, there exist integers a and b such that $aq + bl = 1$. By the irreducibility lemma 8.4.14 (a), $F[r]$ is a field. Then $r = (r^q)^a(r^l)^b = (r^q)^a \rho^b p_l^{-b} \in F[\rho]$. Hence $F[r] = F[\rho]$. $\qquad\square$

Lemma 8.4.14. Let q be a prime, $r \in \mathbb{C}$ a number, and $F \subset \mathbb{C}$ a field containing ε_q and r^q but not r.

(a) **Irreducibility.** The polynomial $t^q - r^q \in F[t]$ is irreducible over F.[13]

(b) **Linear independence.** If $P(r) = 0$ for some polynomial $P \in F[t]$ of degree less than q, then $P = 0$.

(c) **Conjugation.** If $Q \in F[t]$ is a polynomial and $Q(r) = 0$, then $Q(r\varepsilon_q^j) = 0$ for $j = 1, \dots, q-1$.

Proof of (a). The roots of the polynomial $t^q - r^q$ are $r, r\varepsilon_q, r\varepsilon_q^2, \dots, r\varepsilon_q^{q-1}$. Then the constant term of a factor of $t^q - r^q$ is the product of some m of these roots. Since $\varepsilon_q \in F$, we obtain $r^m \in F$. For a proper factor, if it existed, $0 < m < q$. Since q is a prime, $qa + mb = 1$ for some integers a and b. Then $r = (r^q)^a(r^m)^b \in F$, which is a contradiction. $\qquad\square$

Parts (b) and (c) are deduced from (a) similarly to Lemma 8.4.11.

Proof of the Abel–Ruffini Theorem 8.4.12. We may assume that all the degrees of the root extractions are prime numbers, i.e., $F_j = F_{j-1}(r_j)$ for some $r_j \in \mathbb{C}$ such that $r_j^{k_j} \in F_{j-1}$ for some prime k_j. Let us prove the modified statement by induction on s. The base case $s = 0$ is obvious. Let us prove the inductive step.[14]

Take the smallest s for which there is a radical extension $\mathbb{Q}_\varepsilon(\vec{a}) = F_0 \subset \dots \subset F_s \ni x_1$ such that all the $k_1, \dots, k_s$ are prime powers. Call an integer $j \in \{1, \dots, s\}$ *interesting* if $F_j \cap \mathbb{Q}_\varepsilon(\vec{x}) \not\subset F_{j-1}$. By the minimality of the number s, it is interesting. Take the largest $m \le s$ for which there is a radical extension as above such that $m - 1$ is not interesting but $m, m+1, \dots, s$ are.

---

[13] The analogue of the irreducibility lemma without the condition "$\varepsilon_q \in F$" is false for $q > 2$, $F = \mathbb{R}$, and $r = \varepsilon_q$. For example, the condition "$\varepsilon_q \in F$" is omitted in the wonderful book [**Pra07a**, pp. 580–581]. Let us explain this subtle point in more detail. In [**Pra07a**] the statement "$q = p$" on p. 581 (for $p = 2$) means the following: if a quadratic polynomial f is irreducible over a field F containing i but factors over $F[\sqrt[q]{a}]$ for some $a \in F$ and prime q, then $q = 2$. This is incorrect for $f(x) = x^2 + x + 1$, $q = 3$, $a = 1$, and $F = \mathbb{Q}[i]$. The error in the proof in [**Pra07a**] is in the previous sentence: (the correct) Theorem 1 on p. 572 cannot be applied, since perhaps $a = b^q$ for some $b \in F$ (although $\sqrt[q]{a} \notin F$).

[14] This proof is due to I. Gaiday-Turlov and A. Lvov. Like everything in this book, it could have been known earlier.

If $m = 1$, then by the rationalization lemma 8.4.13 we can for $j = 1, 2, \ldots, s$ consecutively replace r_j by ρ_j so that $\mathbb{Q}_\varepsilon(\vec{a}, \rho_1, \ldots, \rho_j) = F_j$. For $j = s$ we obtain the required radical extension.

Now assume that $m > 1$. Since m is interesting, by the rationalization lemma 8.4.13 there is $\rho \in \mathbb{Q}_\varepsilon(\vec{x})$ such that $F_m = F_{m-1}(\rho)$. For the prime k_m we have $\rho^{k_m} \in F_{m-1}$. Since $m-1$ is not interesting, we have $F_{m-1} \cap \mathbb{Q}_\varepsilon(\vec{x}) \subset F_{m-2}$. Hence $\rho^{k_m} \in F_{m-2}$. Thus $F_{m-2}(\rho)$ is a radical extension of F_{m-2} and $F_{m-2}(\rho, r_{m-1}) = F_{m-2}(r_{m-1}, \rho) = F_{m-1}(\rho) = F_m$. So replacement of r_{m-1} and r_m by ρ and r_{m-1} gives a radical extension of F_0 with the same s such that each field except $F_{m-2}(\rho)$ coincides with the corresponding field among $F_0, \ldots, F_s$. For the new extension, m is not interesting but $m+1, \ldots, s$ are. Since $m < s$, this contradicts the maximality of m. $\qquad\square$

4.G. Kronecker's Theorem (4*)

Kronecker's Theorem 8.1.14 (and, thus, the Galois Theorem 8.1.12) follows from the consolidation lemma 8.4.15 and Lemma 8.4.16 (a) below.

For a prime q, a field $F \subset \mathbb{C}$, and a number $r \in \mathbb{C} - F$ such that $r^q \in F$, an extension $F[r]$ of F is called *normal* if $\varepsilon_q \in F$.

Lemma 8.4.15 (Consolidation). If a number $x \in \mathbb{C}$ is expressible in radicals, then there is a tower of normal extensions (from Lemma 8.4.1 (b)) such that for every $k = 1, 2, \ldots, s-1$ either $r_k \in \mathbb{R}$ or $|r_k|^2 \in F_k$.

Lemma 8.4.16. Let q be a prime, $F \subset \mathbb{C}$ a field, and $r \in \mathbb{C} - F$ such that $r^q, \varepsilon_q \in F$.

(a) Suppose that either $r \in \mathbb{R}$ or $|r|^2 \in F$, and let $G \in F[t]$ be a polynomial of prime degree with more than one real root and at least one non-real root. If G is irreducible over F, then G is irreducible over $F[r]$.

(b) **Parametric conjugation.** If $P \in F[x, t]$ and $P(x, r) = 0$ as a polynomial in x, then $P(x, r\varepsilon_q^k) = 0$ as a polynomial in x for every $k = 0, 1, \ldots, q-1$.

Proof of (b). We can replace the polynomial P with its remainder upon division by $t^q - r^q$. Therefore, we can assume that $\deg_t P < q$. In this case, the statement is obtained by applying the linear independence lemma 8.4.14 (b) to the coefficients. $\qquad\square$

Lemma 8.4.17 (Rationality). Let $F \subset \mathbb{C}$ be a field, q an integer, $r \in \mathbb{C}$ with $r^q \in F$, and $H \in F[x, t]$. Then $H(x, r)H(x, \varepsilon_q r) \cdots H(x, \varepsilon_q^{q-1} r) \in F[x]$.

Proof (I. I. Bogdanov). The product $H(x, x_0)H(x, x_1) \cdots H(x, x_{q-1})$ is a symmetric polynomial in $x_0, x_1, \ldots, x_{q-1}$. Thus, it is a polynomial in x and in elementary symmetric polynomials in $x_0, x_1, \ldots, x_{q-1}$. The values of these elementary symmetric polynomials for $x_k = r\varepsilon_q^k$, $k = 0, 1, \ldots, q-1$, are equal to the coefficients of $x^q - r^q$ and are members of F.[15]   □

Proof of Lemma 8.4.16. (a) (We will consider divisibility and irreducibility in $F[r]$ unless otherwise stated.) Suppose to the contrary that G is reducible. Then G has an irreducible divisor in $F[r][x]$. This divisor is the value $H(x, r)$ of a polynomial $H \in F[x, t]$ of degree more than 0 and less than q in t, and of degree less than $\deg G$ in x. So $H(x, r)$ is irreducible and $G(x) = H(x, r)H_1(x, r)$ for some polynomial $H_1 \in F[x, t]$. Let $\varepsilon := \varepsilon_q$. Apply part (b) to $P(x, t) := G(x) - H(x, t)H_1(x, t)$. We see that $G(x)$ is divisible by the polynomial $H(x, r\varepsilon^k)$ for each $k = 0, 1, \ldots, q-1$.

If $H(x, r\varepsilon^k)$ factors for some $k = 0, 1, \ldots, q-1$, then $H(x, r\varepsilon^k) = H_2(x, r)H_3(x, r)$ for some polynomials $H_2, H_3 \in F[x, t]$. Apply (b) to $P(x, t) := H(x, t\varepsilon^k) - H_2(x, t)H_3(x, t)$. This implies that $H(x, r)$ factors, a contradiction. So $H(x, r\varepsilon^k)$ is irreducible for every $k = 0, 1, \ldots, q-1$.

By the linear independence lemma 8.4.14 (b), the polynomials $H(x, r\varepsilon^k)$ for $k = 0, 1, \ldots, q-1$ are distinct. Hence G is divisible by their product. The rationality lemma 8.4.17 asserts that the coefficients of this product lie in F. From this and the irreducibility of G over F, it follows that G equals this product up to a constant multiple $a \in F$. Thus $\deg G = q \deg_x H$. Since $\deg G$ is a prime and $\deg_x H < \deg G$, we have $\deg_x H = 1$ (and $\deg G = q$). So there exist $h_0, \ldots, h_{q-1} \in F$ such that the roots of G are

$$x_k := h_0 + h_1 r\varepsilon^k + \ldots + h_{q-1} r\varepsilon^{k(q-1)}, \quad k = 0, 1, \ldots, q-1.$$

The property $x_k \in \mathbb{R}$ is equivalent to $x_k = \overline{x}_k$. Note that $\overline{\varepsilon_q^k} = \varepsilon_q^{-k}$.

If $r \in \mathbb{R}$, then by the linear independence lemma we have that for every $k = 0, 1, \ldots, q-1$ the condition $x_k = \overline{x}_k$ is equivalent to $h_s \varepsilon^{2sk} = \overline{h}_s$ for every $s = 0, 1, \ldots, q-1$.

Hence $x_k \in \mathbb{R}$ for at most one k. If $r \notin \mathbb{R}$, then $|r|^2 \in F$. Then $\overline{r}^s = \frac{|r|^{2s}}{r^q} r^{qs}$, where $\frac{|r|^{2s}}{r^q} \in F$. Hence the linear independence lemma implies that for every $k = 0, 1, \ldots, q-1$ the condition $x_k = \overline{x}_k$ is equivalent to $h_0 = \overline{h}_0$ and $h_s = \overline{h}_{q-s} \frac{|r|^{2q-2s}}{r^q}$ for every $s = 1, 2, \ldots, q-1$.

These equations do not depend on k. Therefore if one of the numbers $x_0, \ldots, x_{q-1}$ is real, then they are all real, which is a contradiction.[16]   □

---

[15]*Another proof.* By the linear independence lemma 8.4.14 (b), this product can be uniquely represented in the form

$$a_0(x) + a_1(x)r + \ldots + a_{q-1}(x)r^{q-1} \quad \text{for some } a_k \in F[x].$$

The product is invariant under the substitution $r \to r\varepsilon$, which is well-defined by the linear independence lemma. Using this lemma again, we see that $a_k(x) = a_k(x)\varepsilon^k \in F[x]$ for every $k = 1, 2, \ldots, q-1$. Hence $a_k(x) = 0$ for every $k = 1, 2, \ldots, q-1$. Thus the product equals $a_0(x) \in F[x]$.

[16] The two cases to be investigated at the end of this proof are slightly different from the cases analyzed at the end of the proof in the article [**Tik03**]. The beginning of the second column on p. 14 in [**Tik03**] actually uses the fact that $\rho \in \mathbb{R}$, but this is not true without additional results, such as the condensation lemma 8.4.15.

Proof of the consolidation lemma 8.4.15. Let us show by downward induction on q that *from an arbitrary tower of extensions one can obtain a tower of extensions for which* $\varepsilon_{q_k} \in F_k$ *for every* $k = 1, 2, \ldots, s-1$ *such that* $q_k > q$. Then for $q = 1$ we obtain a tower of normal extensions. The base case is $q = \max_k q_k$; in this case there is nothing to prove. To prove the inductive step, consider the least k such that $q_k = q$. If such a k does not exist, then the inductive step is obvious: Paste "between" F_{k-1} and F_k, "getting ε_q by extracting only roots of degree less than q" obtained from Gauss's degree-lowering theorem 8.1.15, increasing "by necessity" the fields $F_k, \ldots, F_s$. More precisely, consider the tower

$$F_{k-1} \subset G_1 \subset G_2 \subset \ldots \subset G_m \subset F_{k-1}[\varepsilon_q]$$

from Gauss's degree-lowering theorem 8.1.15. Replace the subtower $F_{k-1} \subset F_k \subset \ldots \subset F_s$ with the subtower

$$F_{k-1} \subset G_1 \subset G_2 \subset \ldots \subset G_m \subset F_{k-1}[\varepsilon_q] \subset F_k[\varepsilon_q] \subset \ldots \subset F_s[\varepsilon_q].$$

Then, whenever possible, replace every extraction of a root of a composite degree ab by extraction of roots of ath and bth degrees. The condition "$\varepsilon_{q_k} \in F_k$ for every $k = 1, 2, \ldots, s-1$ such that $q_k \geq q$" is preserved, because if $\varepsilon_{ab} \in F_k$, then $\varepsilon_a \in F_k$ and $\varepsilon_b \in F_k$. In the new subtower, replace the repeated copies of the same field by a single field. The inductive step is proved.

Let us show by downward induction on l that *from an arbitrary tower of normal extensions one can obtain a tower of normal extensions such that for every* $k \leq s - l$,

$$\overline{F}_k = F_k \quad \text{and either} \quad r_k \in \mathbb{R} \quad \text{or} \quad |r_k|^2 \in F_k.$$

Then for $l = 0$ we obtain the lemma. The base case is $l = s - 1$, for which there is nothing to prove. Let us prove the inductive step. (If $r_k \in \mathbb{R}$, then the inductive step is obvious, but the following argument also works.) Since $F_k = \overline{F_k}$ and $r_k^{q_k} \in F_k$, we obtain $|r_k|^{2q_k} = r_k^{q_k} \overline{r_k^{q_k}} \in F_k$. So $F_k[|r_k|^2] = F_k[\sqrt[q_k]{|r_k|^{2q_k}}]$, where we choose the real value of the root. Replace the subtower $F_k \subset F_{k+1} \subset \ldots \subset F_s$ by the subtower

$$F_k \subset F_k[|r_k|^2] \subset F_k[r_k, \overline{r}_k] = F_{k+1}[\overline{r}_k] \subset \ldots \subset F_s[\overline{r}_k].$$

Clearly, normality is preserved under this substitution. In the new subtower, replace repeated copies of the same field by one field. After that, apply the inductive hypothesis. The inductive step is proved. $\square$

Remark 8.4.18. (a) The difference between the proofs of Lemmas 8.4.16 (a) and 8.4.11 (a) lies in

• "complexification"—r^q and the coefficients of the polynomial H may be complex;

• the necessity to prove the existence of a root in $F[r]$ of a polynomial irreducible over F and reducible over $F[r]$ (or to assume that the polynomial has a root in $F[r]$ and does not have a root in F, and to prove its irreducibility over F, which is less convenient).

(b) Kronecker's Theorem 8.1.14 can be proved using Lemma 8.4.20 (on the equivalence of irreducibility over F and over $F[\varepsilon_q]$). Then there is no need to require normality in the consolidation lemma 8.4.15 using Gauss's degree-lowering theorem 8.1.15. The details are similar to the proof of the real analogue 8.1.10 of Kronecker's Theorem; see 4.H. But to prove this analogue, requiring normality in the consolidation lemma (instead of using Lemma 8.4.20), is not possible.

4.H. The real analogue of Kronecker's Theorem (4*)

Theorem 8.1.10 is implied by the tower of extensions lemma 8.4.1 (b) and the following.

Lemma 8.4.19. Let q be a prime, $F \subset \mathbb{R}$ a field and $r \in \mathbb{R} - F$ with $r^q \in F$.

(a) If a polynomial $G \in F[t]$ of prime degree has more than one real root and is irreducible over F, then G is irreducible over $F[r]$.

(b) **Parametric conjugation.** If $P \in F[x, t]$ and $P(x, r) = 0$ as a polynomial in x, then $P(x, r\varepsilon_q^k) = 0$ as a polynomial in x for $k = 0, 1, \ldots, q - 1$.

Part (a) is interesting even for $F = \mathbb{Q}$ (although insolvability with a single root extraction can be proved without it) and is true even for $F \subset \mathbb{C}$.

The proof of (b) is similar to the proof of Lemma 8.4.16 (a). Instead of Lemma 8.4.14 (b), we apply Lemma 8.4.11 (c). Note that Lemma 8.4.11 (c) is implied by Lemma 8.4.14 (b) and the following.

Lemma 8.4.20. If a polynomial of prime degree p is irreducible over the field $F \subset \mathbb{C}$ and is reducible over $F[\varepsilon_q]$ for some q, then $q > p$.

Proof of Lemma 8.4.19 (a). Following the proof of Lemma 8.4.16 (a), we repeat the first three paragraphs with the replacement of F with $F[\varepsilon_q]$ (and therefore $F[r]$ with $F[\varepsilon_q][r]$). In the third paragraph, since G is divisible by the product, $\deg G \geq q$. From that and from Lemma 8.4.20, it follows that G is irreducible over $F[\varepsilon_q]$. Finally, consider the case $r \in \mathbb{R}$ and do not consider the case $r \notin \mathbb{R}$. □

It remains to prove Lemma 8.4.20.

Lemma 8.4.21. For $\alpha, \beta \in \mathbb{C}$ and a field $F \subset \mathbb{C}$, define $[\alpha : \beta]$ to be the degree of a polynomial irreducible over $F[\beta]$ with root α, if such a polynomial exists, and let $[\alpha : \beta] = \infty$ if it does not. Then $[\alpha : 1][\beta : \alpha] = [\beta : 1][\alpha : \beta]$ for any $\alpha, \beta \in \mathbb{C}$.

Proof (sketch). We have already seen that irreducibility is connected with linear independence. This motivates the following approach. For fields $K \subset L$ define the *dimension* $\dim_K L$ *of a field L over a field K* to be the smallest s for which there exist s elements $l_1, \ldots, l_s \in L$ such that for any $l \in L$ there exist $k_1, \ldots, k_s \in K$ such that $l = k_1 l_1 + \ldots + k_s l_s$. For example, $\dim_{\mathbb{Q}} \mathbb{Q}[\sqrt[5]{3}] = 5$.

The lemma is a consequence of the following two statements. The proofs are left to the reader as exercises.

(a) If $\alpha \in \mathbb{C}$ is a root of a polynomial P that is irreducible over the number field F, then $\dim_F F[\alpha] = \deg P$.

(b) For any fields $K \subset L \subset M$ we have $\dim_K M = \dim_L M \cdot \dim_K L$. □

Proof of Lemma 8.4.20. Let $\beta \in \mathbb{C}$ be any root of the given polynomial. Lemma 8.4.21 implies $[\varepsilon_q : 1][\beta : \varepsilon_q] = [\beta : 1][\varepsilon_q : \beta]$. Since the polynomial is irreducible over F, we have $[\beta : 1] = p$. Since the polynomial is reducible over $F[\varepsilon_q]$, $[\beta : \varepsilon_q] < p$. We have $[\varepsilon_q : 1] < q$. Since p is prime, $[\varepsilon_q : 1]$ is divisible by p. Therefore, $q > p$. □

8.4.22. Is $\cos(2\pi/n)$ expressible in real radicals for
    (a) $n = 11$;   (b)* $n = 13$?

Bibliography

[ABG+] D. Akhtyamov, I. Bogdanov, A. Glebov, A. Skopenkov, E. Streltsova, and A. Zykin, *Solving equations using one radical*, `http://www.turgor.ru/lktg/2015/4/index.htm` (Russian).

[Akh] D. Akhtyamov, and I. Bogdanov, *Solvability of cubic and quartic equations using one radical*, `arXiv:1411.4990` (Russian).

[Ale04] V. B. Alekseev, *Abel's theorem in problems and solutions*, Springer-Verlag, 2004.

[Arn16a] V. I. Arnold, *Continued fractions*, in *Lectures and problems: A gift to young mathematicians*, American Mathematical Society, 2016.

[Arn16b] ———, *Problems for children 5 to 15 years old*, in *Lectures and problems: A gift to young mathematicians*, American Mathematical Society, 2016.

[AS03] J. P. Allouche and J. Shallit, *Automatic sequences: Theory, applications, generalizations*, Cambridge University Press, 2003.

[AS16a] N. Alon and J. Spencer, *The probabilistic method*, John Wiley & Sons, 2016.

[AS16b] T. Andreescu and M. Saul, *Algebraic inequalities: New vistas*, Mathematical Sciences Research Institute and American Mathematical Society, 2016.

[Ay] R. G. Ayoub, *On the nonsolvability of the general polynomial*, Amer. Math. Monthly **89** (1982), no. 6, 397–401.

[AZ04] M. Aigner and G. Ziegler, *Proofs from the book*, Springer-Verlag, 2004.

[Bar03] E. J. Barbeau, *Polynomials*, Springer-Verlag, 2003.

[BB65] E. F. Beckenbach and R. Bellman, *Inequalities*, Springer-Verlag, 1965.

[Ben88] A. Bendukidze, *Fermat searches for extrema*, Kvant (1988), no. 10, 45–48 (Russian).

[Ber10] J. Bergen, *A concrete approach to abstract algebra: From the integers to the unsolvability of the quintic*, Academic Press, 2010.

[Bew06] Jörg Bewersdorff, *Galois theory for beginners: A historical perspective*, American Mathematical Society, 2006.

[BK13] Yu. Burda and L. Kadets, *The heptadecagon and Gauss's reciprocity law*, Matematicheskoye Prosveshcheniye (Mathematical Enlightenment) **17** (2013) (Russian).

[Bro] J. Brown, *Abel and the unsolvability of the quintic*, `http://www.math.caltech.edu/~jimlb/abel.pdf`.

[BSe] A. D. Blinkov and A. V. Shapovalov (eds.), *School mathematical circles* (book series; in Russian).

[Cev] *Ceva's theorem*, `https://artofproblemsolving.com/wiki/index.php/title=Ceva%27s_Theorem`.

[CG67] H. S. M. Coxeter and S. L. Greitzer, *Geometry revisited*, MMA Press, 1967.

[Che34] N. Chebotarev, *Foundations of Galois theory, Part 1*, Gostekhizdat, 1934 (Russian).

[Che16] Evan Chen, *Euclidean geometry in mathematical olympiads*, The Mathematical Association of America, 2016.

[CR96] R. Courant and H. Robbins, *What is mathematics?*, Oxford University Press, 1996.

[DMSF] A. V Doledenok, A. B Menshikov, A. S Semchenkov, and M. A. Fadin, *The pqr method*, http://www.turgor.ru/lktg/2016/3/index.htm (Russian).

[Dob04] V. N. Dobrovolskaya, *Incomplete sums of fractional parts*, Chebyshev Collection **5** (2004), no. 2, 42–48 (Russian).

[Dör13] H. Dörrie, *100 great problems of elementary mathematics: Their history and solution*, Dover, 2013.

[DY85] S. Dvoryaninov and E. Yasinovy, *How are symmetric inequalities obtained?*, Kvant (1985), no. 7, 33–36 (Russian).

[ECG+] A. Enne, A. Chilikov, A. Glebov, A. Skopenkov, and B. Vukorepa, *Algorithms for solving algebraic equations*, https://www.turgor.ru/lktg/2018/5/index.html (Russian).

[Edw97] H. M. Edwards, *Galois theory*, Springer-Verlag, 1997.

[Edw09] _____, *The construction of solvable polynomials*, Bull. Amer. Math. Soc. **46** (2009), 397–411.

[Est] A. Esterov, *Galois theory for general systems of polynomial equations*, arXiv:1801.08260.

[FBKY11a] R. Fedorov, A. Belov, A. Kovaldzhi, and I. Yashchenko (eds.), *Moscow mathematical olympiads 1993–1999*, Mathematical Sciences Research Institute and American Mathematical Society, 2011.

[FBKY11b] R. Fedorov, A. Belov, A. Kovaldzhi, and I. Yashchenko (eds.), *Moscow mathematical olympiads 2000–2005*, Mathematical Sciences Research Institute and American Mathematical Society, 2011.

[Fel83] N. Feldman, *Algebraic and transcendental numbers*, Kvant (1983), no. 7, 2–7 (Russian).

[FK91] Dmitry Fomin and Alexey Kirichenko, *Leningrad mathematical olympiads 1987–1991*, MathPro Press, 1991.

[FT07] D. Fuchs and S. Tabachnikov, *Mathematical omnibus*, American Mathematical Society, 2007.

[Gal80] A. Galochkin, *On the degree of transcendence of values of functions satisfying certain functional equations*, Mathematical Notes **27** (1980), no. 2, 175–183 (Russian).

[Gau] Carl Friedrich Gauss, *Werke*, https://gdz.sub.uni-goettingen.de/id/PPN236010751.

[Ger99] M. L. Gerver, *On partitioning of sets into parts of smaller diameter: Theorems and counterexamples*, Matematicheskoye Prosveshcheniye (Mathematical Enlightenment) **3** (1999), 168–183 (Russian).

[GIF94] S. A. Genkin, I. V. Itenberg, and D. V. Fomin, *Leningrad mathematical circles*, Kirov, 1994 (Russian).

[Gin72] S. Gindikin, *Gauss's debut*, Kvant (1972), no. 1, 2–11 (Russian).

[Gin76] _____, *The great art*, Kvant (1976), no. 9, 2–10 (Russian).

[Gin07] _____, *Tales of mathematicians and physicists*, Springer-Verlag, 2007.

[GKP94] R. Graham, D. Knuth, and O. Patashnik, *Concrete mathematics*, Addison-Wesley, 1994.

[Go09] M. Gorelov, *Inequalities and parallel trasport*, Kvant (2010), no. 2, 41–44.

[Gor10] _____, *The usefulness of graphs*, Kvant (2010), no. 4, 44–47 (Russian).

[Had78] C. R. Hadlock, *Field theory and its classical problems*, Carus Mathematical Monographs, no. 19, The Mathematical Association of America, 1978.

[HLP67] G. H. Hardy, J. E. Littlewood, and G. Pólya, *Inequalities*, Cambridge University Press, 1967.

[IBL]   *Inquiry-based learning*, https://en.wikipedia.org/wiki/Inquiry-based_learning.

[ISO]   *Isogonal conjugate*, https://en.wikipedia.org/wiki/Isogonal_conjugate.

[Kan]   A. L. Kanunnikov, *Introductory Galois theory: Solvability of algebraic equations by radicals*, http://www.mathnet.ru/conf1015 (Russian).

[KBK08]   A. Ya. Kanel-Belov and A. K. Kovalji, *How to solve non-standard problems*, Moscow Center for Continuous Mathematical Education, 2008 (Russian).

[KBK15]   ———, *Mathematics classes: Leaflets and dialogue*, Matematicheskoye Prosveshcheniye (Mathematical Enlightenment) **19** (2015), 206–233 (Russian).

[Kho13]   A. G. Khovansky, *Constructions by compass and straightedge*, Matematicheskoye Prosveshcheniye (Mathematical Enlightenment) **17** (2013) (Russian).

[Kir]   V. A. Kirichenko, *Constructions by compass and straightedge and Galois theory*, http://www.mccme.ru//dubna/2005/courses/kirichenko.html (Russian).

[Kir77]   A. A. Kirillov, *On regular polygons, the Euler function, and Fermat numbers*, Kvant (1977), no. 7, 2–9 (Russian).

[Kog]   E. S. Kogan, *Set complexity for the construction of a regular polygon*, arXiv:1711.05807 (Russian).

[Kol01]   V. A. Kolosov, *Theorems and problems in algebra, number theory and combinatorics*, Helios, 2001 (Russian).

[KPV02]   Kiran S. Kedlaya, Bjorn Poonen, and Ravi Vakil (eds.), *The William Lowell Putnam Mathematical Competition*, The Mathematical Association of America, 2002.

[KS08]   P. Kozlov and A. Skopenkov, *A la recherche de l'algèbre perdue: du cote de chez Gauss*, Matematicheskoye Prosveshcheniye (Mathematical Enlightenment) **12** (2008), 127–144, arXiv:0804.4357v1.

[Kur63a]   Jozsef Kurschak, *Hungarian problem book I*, The Mathematical Association of America, 1963.

[Kur63b]   ———, *Hungarian problem book II*, The Mathematical Association of America, 1963.

[L]   L. Lerner, *Galois theory without abstract algebra*, arXiv:1108.4593.

[Liu01]   Andy Liu (ed.), *Hungarian problem book III*, The Mathematical Association of America, 2001.

[LKT]   *Summer conferences of the Tournament of Towns*, http://www.turgor.ru/en/lktg/index.php.

[M]   *Moscow Mathematical Conference of High School Students*, http://www.mccme.ru/mmks/index_eng.htm.

[Mah29]   K. Mahler, *Arithmetische eigenschaften der lösungen einer klasse von funktional-gleichungen*, Mathematische Annalen **1** (1929), 342–366.

[Man63]   Yu. I. Manin, *On the solvability of construction problems by compass and straight edge*, in *Encyclopedia of elementary mathematics, Book 4 (Geometry)* (P. S. Aleksandrov, A. I. Markushevich, and A. Ya. Khinchin, eds.), Fizmatgiz, 1963 (Russian).

[Mey]   D. Meyer, *Starter pack*, http://blog.mrmeyer.com/starter-pack.

[MO09]   A. W. Marshall and I. Olkin, *Inequalities: Theory of majorization and its applications*, Springer-Verlag, 2009.

[Nil94]   A. Nilli, *On Borsuk's problem*, Contemp. Math. (1994), no. 178, 209–210.

[Nis96]   K. Nishioka, *Mahler functions and transcendence*, Lecture Notes in Math, vol. 1631, Springer-Verlag, 1996.

[Pes04]   P. Pesic, *Abel's proof*, The MIT Press, 2004.

[Pla12]   Plato, *Phaedo*, Kindle edition, Amazon Digital Services, 2012.

[Pon84]   L. Pontryagin, *The cubic parabola*, Kvant (1984), no. 3, 10–14, 32.

[Pos78] M. M. Postnikov, *Fermat's theorem: Introduction to the theory of algebraic numbers*, Nauka, 1978 (Russian).

[Pos14] _____, *Fundamentals of Galois theory*, Pergamon Press, 2014.

[PR98] B. Poonen and M. Rubinstein, *The number of intersection points made by the diagonals of a regular polygon*, SIAM J. Disc. Math. **11** (1998), 133–156, arXiv:9508209.

[Pra07a] V. V. Prasolov, *Problems in algebra, arithmetic, and analysis*, Moscow Center for Continuous Mathematical Education, 2007 (Russian).

[Pra07b] _____, *Problems in plane geometry*, Moscow Center for Continuous Mathematical Education, 2007 (Russian).

[PS97] V. V. Prasolov and Y. P. Solovyev, *Elliptic functions and elliptic integrals*, American Mathematical Society, 1997.

[PS04] G. Pólya and G. Szegő, *Problems and theorems in analysis*, Springer-Verlag, 2004.

[RSG+16] A. M. Raĭgorodskiĭ, A. B. Skopenkov, A. A Glibichuk, A. B. Dajnyak, D. G. Ilyinsky, A. B. Kupavsky, and A. A. Chernov, *Elements of discrete mathematics in problems*, Moscow Center for Continuous Mathematical Education, 2016 (Russian).

[Rai04] A. M. Raigorodskii, *The Borsuk partition problem: The seventieth anniversary*, Math. Intelligencer **26** (2004), no. 3, 4–12.

[RMP] *Ross mathematics program*, http://u.osu.edu/rossmath.

[Ros95] M. I. Rosen, *Niels Hendrik Abel and equations of the fifth degree*, American Mathematical Monthly **102** (1995), no. 6, 495–505.

[Saf] A. Safin, *An algorithm for the construction of regular polygons by compass and ruler*, http://www.mccme.ru/mmks/dec08/Safin.pdf (Russian).

[SC93] D. O. Shklyarsky and N. N. Chentzov (eds.), *The USSR olympiad problem book: Selected problems and theorems of elementary mathematics*, Dover, 1993.

[Sgi09] A. Sgibnev, *How to approximately calculate the roots to solve cubic equations*, Potentsial (2009), no. 12 (Russian).

[Siv67] I. C. Sivashinsky, *Inequalities in problems*, Nauka, 1967 (Russian).

[Skoa] A. Skopenkov, *Basic embeddings and Hilbert's 13th problem*, arXiv:1003.1586.

[Skob] A. Skopenkov, *The Borsuk problem*, Quantum **7** (1996), no. 1, 16–21,63.

[Skoc] A. Skopenkov, *Examples of transcendental numbers*, http://www.turgor.ru/lktg/2002/problem5.ru (Russian).

[Skod] _____, *A short elementary proof of the insolvability of the equation of degree 5*, http://arxiv.org/abs/1508.03317.

[Sko09] A. B. Skopenkov, *Fundamentals of differential geometry through interesting problems*, Moscow Center for Continuous Mathematical Education, 2009, arXiv:0801.1568 (Russian).

[Sko10] A. Skopenkov, *Basic embeddings and Hilbert's 13th problem*, Matematicheskoye Prosveshcheniye (Mathematical Enlightenment) **14** (2010), 143–174, arXiv:1001.4011 (Russian).

[Sko11] _____, *A simple proof of the Abel-Ruffini theorem on the unsolvability of equations in radicals*, Matematicheskoye Prosveshcheniye (Mathematical Enlightenment) **15** (2011), 113–126, arXiv:1102.2100 (Russian).

[Sko21] M. B. Skopenkov and A. A. Zaslavsky, *Mathematics Through Problems. Part II: Combinatorics*, AMS, Providence, Rhode Island, forthcoming in 2021.

[Sop] J. Soprunova, *Pick's formula*, http://www.math.kent.edu/~soprunova/64091s15/Pick.

[Ste04] J. Michael Steele, *The Cauchy-Schwarz master class: An introduction to the art of mathematical inequalities*, Cambridge University Press, 2004.

[Sti94] J. Stillwell, *Galois theory for beginners*, American Mathematical Monthly **101** (1994), 22–27.

[SB78] S. Strashevich and E. Brovkin, *Polish mathematical olympiads*, "Mir", Moscow, 1978.

[Suz18] D. T. Suzuki, *An introduction to Zen Buddhism*, Kindle edition, Amazon Digital Services, 2018.

[Tab88] S. Tabachnikov, *The geometry of equations*, Kvant (1988), no. 10, 10–16.

[Tab99] S. Tabachnikov (ed.), *Kvant selecta: Algebra and analysis I*, vol. 2, American Mathematical Society, 1999.

[Tab01] S. Tabachnikov (ed.), *Kvant selecta: Combinatorics I*, vol. 1, American Mathematical Society, 2001.

[Tik94] V. M. Tikhomirov, *Chebyshev's theorem on the distribution of prime numbers*, Kvant (1994), no. 6, 12–13 (Russian).

[Tik03] V. M. Tikhomirov, *Abel and his great theorem*, Kvant (2003), no. 1, 1–15 (Russian).

[V+15] N. Ya. Vilenkin et al., *Algebra and mathematical analysis: Textbook for 11th grade*, Mnemozina, 2015 (Russian).

[Vag80] N. Vaguten, *Conjugation*, Kvant (1980), no. 2, 26–32 (Russian).

[Vas87] N. B. Vasiliev (ed.), *Mathematical olympiad correspondence exams*, Nauka, 1987 (Russian).

[Vin80] E. B. Vinberg, *The algebra of polynomials*, Matematicheskoye Prosveshcheniye (Mathematical Enlightenment) (1980) (Russian).

[VINK10] O. Ya. Viro, O. A. Ivanov, N. Yu. Netsvetaev, and V. M. Kharlamov, *Elementary topology: Problem textbook* American Mathematical Society, 2008.

[VSY17] A. Volostnov, A. Skopenkov, and Yu. Yarovikov, *On recurrence relations*, Matematicheskoye Prosveshcheniye (Mathematical Enlightenment) **21** (2017) (Russian).

[Vyg67] M. Ya. Vygodsky, *Arithmetic and algebra in the ancient world*, Nauka, 1967 (Russian).

[Yan] Dian Yang, *An elementary proof of Borsuk's theorem*, http://arxiv.org/abs/1010.1990.

[Yu70] *Mathematics in the XVII century* in *History of mathematics* (A. P. Yushkevich, ed.), vol. 2, Nauka, 1970, http://ilib.mccme.ru/djvu/istoria/istmat2.htm (Russian).

[Zor15] V. A. Zorich, *Mathematical analysis, Part I*, Moscow Center for Continuous Mathematical Education, 2015 (Russian).

[Zvo12] A. K. Zvonkin, *Math from three to seven: The story of a mathematical circle for preschoolers*, Mathematical Sciences Research Institute and American Mathematical Society, 2012.

Index